STUDIES IN GEOPHYSICS

Fundamental Research on Estuaries: The Importance of an Interdisciplinary Approach

Panel on Estuarine Research Perspectives

Geophysics Study Committee

Geophysics Research Board

Commission on Physical Sciences, Mathematics, and Resources

National Research Council

NATIONAL ACADEMY PRESS
Washington, D.C.   1983

The Geophysics Study Committee is pleased to acknowledge the support of the National Science Foundation (EAR-79-02458), the Defense Advanced Research Projects Agency, the National Aeronautics and Space Administration, the National Oceanic and Atmospheric Administration (NA80AAA04492), the U.S. Geological Survey, and the Department of Energy (DE-FGO2-80ER10757 and DE-FGO2-81EV10452) for the conduct of this study.

International Standard Book Number 0-309-03378-0

Library of Congress Catalog Card Number 83-60962

Available from

NATIONAL ACADEMY PRESS
2101 Constitution Avenue, NW
Washington, DC 20418

Printed in the United States of America

PANEL ON
ESTUARINE RESEARCH PERSPECTIVES

CHARLES B. OFFICER, Dartmouth College, *Cochairman*
L. EUGENE CRONIN, Chesapeake Research Consortium,
    Cochairman
ROBERT B. BIGGS, University of Delaware
JAMES M. COLEMAN, Louisiana State University
DONALD V. HANSEN, National Oceanic and Atmospheric
    Administration
ALLAN J. MEARNS, National Oceanic and Atmospheric
    Administration
SCOTT W. NIXON, University of Rhode Island
DAVID H. PETERSON, U.S. Geological Survey
DONALD W. PRITCHARD, State University of New York, Stony
    Brook
MAURICE RATTRAY, University of Washington
JOHN H. RYTHER, Harbor Branch Institution
KARL K. TUREKIAN, Yale University
ROBERT E. WILSON, State University of New York, Stony
    Brook
HERBERT L. WINDOM, Skidaway Institute of Oceanography

GEOPHYSICS
STUDY COMMITTEE*

ARTHUR E. MAXWELL, The University of Texas, Austin, *Chairman*
CHARLES L. DRAKE, Dartmouth College, *Chairman* (1982)
LOUIS J. BATTAN, University of Arizona, *Vice Chairman* (1982)
JOHN D. BREDEHOEFT, U.S. Geological Survey (1982)
COLIN BULL, The Ohio State University
ALLAN V. COX, Stanford University (1982)
JOHN C. CROWELL, University of California, Santa Barbara
NICHOLAS C. MATALAS, U.S. Geological Survey
J. MURRAY MITCHELL, National Oceanic and Atmospheric
    Administration
V. RAMA MURTHY, University of Minnesota
HUGH ODISHAW, University of Arizona (1982)
CHARLES B. OFFICER, Dartmouth College (1982)
RAYMOND G. ROBLE, National Center for Atmospheric Research
FERRIS WEBSTER, University of Delaware

Liaison Representatives
LEONARD E. JOHNSON, National Science Foundation
BRUCE B. HANSHAW, U.S. Geological Survey
GEORGE A. KOLSTAD, Department of Energy
NED A. OSTENSO, National Oceanic and Atmospheric
    Administration
WILLIAM RANEY, National Aeronautics and Space Administration
CARL F. ROMNEY, Defense Advanced Research Projects Agency

Staff
THOMAS M. USSELMAN

*Expiration of term of former members who participated in
guiding the study is shown in parenthesis.

iv

GEOPHYSICS
RESEARCH BOARD

STUDIES IN GEOPHYSICS*

ENERGY AND CLIMATE
  Roger R. Revelle, *panel chairman*, 1977, 158 pp.

CLIMATE, CLIMATIC CHANGE, AND WATER SUPPLY
  James R. Wallis, *panel chairman*, 1977, 132 pp.

ESTUARIES, GEOPHYSICS, AND THE ENVIRONMENT
  Charles B. Officer, *panel chairman*, 1977, 127 pp.

THE UPPER ATMOSPHERE AND MAGNETOSPHERE
  Francis S. Johnson, *panel chairman*, 1977, 169 pp.

GEOPHYSICAL PREDICTIONS
  Helmut E. Landsberg, *panel chairman*, 1978, 215 pp.

IMPACT OF TECHNOLOGY ON GEOPHYSICS
  Homer E. Newell, *panel chairman*, 1979, 121 pp.

CONTINENTAL TECTONICS
  B. Clark Burchfiel, Jack E. Oliver, and Leon T. Silver,
  panel cochairmen, 1980, 197 pp.

MINERAL RESOURCES: GENETIC UNDERSTANDING FOR PRACTICAL
APPLICATIONS
  Paul B. Barton, Jr., *panel chairman*, 1981, 118 pp.

SCIENTIFIC BASIS OF WATER RESOURCE MANAGEMENT
  Myron B. Fiering, *panel chairman*, 1982, 127 pp.

SOLAR VARIABILITY, WEATHER, AND CLIMATE
  John A. Eddy, *panel chairman*, 1982, 106 pp.

CLIMATE IN EARTH HISTORY
  Wolfgang H. Berger and John C. Crowell, *panel
  cochairmen*, 1982, 197 pp.

FUNDAMENTAL RESEARCH ON ESTUARIES: THE IMPORTANCE OF AN
INTERDISCIPLINARY APPROACH
  Charles B. Officer and L. Eugene Cronin, *panel
  cochairmen*, 1983, 79 pp.

*Published to date.

In 1974 the Geophysics Research Board completed a plan, subsequently approved by the Committee on Science and Public Policy of the National Academy of Sciences, for a series of studies to be carried out on various subjects related to geophysics. The Geophysics Study Committee was established to provide guidance in the conduct of the studies.

One purpose of the studies is to provide assessments from the scientific community to aid policymakers in decisions on societal problems that involve geophysics. An important part of such an assessment is an evaluation of the adequacy of present geophysical knowledge and the appropriateness of present research programs to provide information required for those decisions.

This study was motivated by the recommendation in a previous report (*Estuaries, Geophysics, and the Environment*, 1977) in the Studies in Geophysics series that "an overall review of the national efforts in estuarine research should be conducted." This study examines the need for basic research to understand estuarine processes. The major unanswered questions relate to the interrelationships of estuarine circulation, biota, geology, and chemistry, where an interdisciplinary coordinated effort will be necessary.

The Panel met and agreed that the basic research problems to be assessed are those estuarine processes that require an interdisciplinary solution and that the questions should be tractable within the next ten years. The Panel selected four broad areas of research and established subpanels, which included active participation of additional experts from the scientific community, to address each of the areas. The areas were organized around a traditional disciplinary facet of estuaries but focused on the interdisciplinary needs.

CONTENTS

_____

FUNDAMENTAL RESEARCH ON ESTUARIES:
THE IMPORTANCE OF AN
INTERDISCIPLINARY APPROACH

1 OVERVIEW AND RECOMMENDATIONS

The 850 estuaries of the nation represent, in proportion
to their size, one of the most valuable portions of our
environment. Knowledge of the complex interactions of the
physical, chemical, geologic, and biological processes in
estuaries is fundamental to our understanding of behavior
in estuarine systems. However, these interactions have
not been studied comprehensively. Most estuarine research
has been conducted along specific disciplinary lines.
This report addresses the need for interdisciplinary
research on the entire estuarine system of which three
major components that need to be integrated are the
following:

1. Environmental effects on estuarine biota,
2. Circulation and mixing in estuarine and coastal
   waters, and
3. Suspended and dissolved matter in estuaries.

The interplay of these major components is significant in
almost all estuarine processes, however, details within
the confines of the interactive processes can differ among
specific estuaries or in specific portions of a single
estuary. The understanding of these processes can be
applied to improve management decisions and efforts
related to the entire estuarine environment--helping in
determining their best use and survival as a productive
and valuable resource. Most important in the management
of estuarine systems is the development of predictive
models,[1] which are both process-oriented (causal) and
empirically based. Much of the past estuarine research
has been based on "after the fact" or present condition
analysis of the system that provides an empirical data
base. However, to develop and verify useful predictive

3

models, more attention needs to be focused on understanding the processes involved.

In the following chapters, teams of appropriate experts comment on recent significant research within the major estuarine components, explain the need for further progress, note the necessity for integrating the other components, and identify important tractable research topics.

USES OF ESTUARIES

Estuaries are semienclosed bodies of water--bays, sounds, inlets, fjords, and lagoons--that have free connection with the open sea and contain water from land drainage and seawater. Each estuary is unique in its totality, but there are primary characteristics that permit judicious transfer of at least some of what is learned in one estuary to some or all of the others. All estuaries contain chemical gradients between their saline and fresh sources of water input. Frequently, there are other gradients found in estuaries--temperature, sediment burden, and dissolved materials. Most estuaries are subjected to relatively violent environmental changes with fluctuations in riverine inflow, seasonal temperatures, and influences of meteorological events. All estuaries are recipients of the composite chemical burden of the waters from land-- waters from urban centers and from all other parts of the contributing watersheds. Most estuaries are or have been extraordinarily productive of biological crops, many of which are useful to man. Each estuary is subject to effects of changes in water quality in those adjacent ocean waters that provide about half of the water in the average estuary.

The value to humans of the estuaries in the United States is high. A major fraction of commercial shipping begins or ends in them. The shoreline is attractive to millions of people for residential sites, for many water-related industries, and for use in recreation (typically on the order of hundreds of millions of visitor-days per year). Boating, swimming, surfing, sunning, hunting, fishing, and nonspecific personal enjoyment are intensive at many sites. Sand, gravel, oil, gas, shell, and commercially useful quantities of many chemicals are mined from the floors of estuaries. The generation of electricity often involves use of estuaries for the release of waste heat and other materials and for transportation. The

military agencies of the nation use bays for transportation, firing ranges, research on vessels and equipment,
storage of vessels, vessel construction and maintenance,
and education in naval and air operations. Estuaries are
excellent sites for the training of scientists and the
conduct of research; laboratories exist in every coastal
state for this kind of activity.

The biological harvest of species dependent on estuaries includes more than 70 percent of the total national
landings by all commercial fisheries and 65 percent of the
national recreational catch in marine waters. In more
specific terms, estuaries were essential for the production in 1980 of 5186 million pounds of commercial species
with a dockside value of $1790 million and a capitalized
annual value of over $35 billion.[2] The recreational
fishing dependent on estuaries involves annual expendi-
tures of over $1 billion and a catch of about 190 million
fish.

The other major use of estuaries is for the placement
--misleadingly referred to as disposal--of wastes. These
bodies of water receive partially treated human wastes,
sediment and chemicals from agricultural activity and
urban runoff, and an enormous array of industrial products
and by-products.

Every one of these uses is predicted to increase, and
each of them can substantially affect the primary components and processes of estuaries so as to reduce the
resource capacity of the system.[3] The legislated but
unaccomplished water-quality goal that all pollutants be
eliminated has an uncertain future, and it must be
assumed, for the near term at least, that wastes will
continue to be introduced.

COMPLEXITIES OF ESTUARINE PROCESSES

Each of the principal components--physical, geologic,
chemical, and biological processes--of an estuarine system
is complex in itself and intimately interactive with the
others. For example, the interactions of population den-
sities in the biological community have been the subject
of a significant amount of research, but when the topic
is considered in light of changes in circulation patterns,
water chemistry,[4] or meteorological effects, the subject
becomes increasingly more complex and dependent on understanding the other components. Figure 1.1 illustrates
some of the complexities of external factors that affect

6

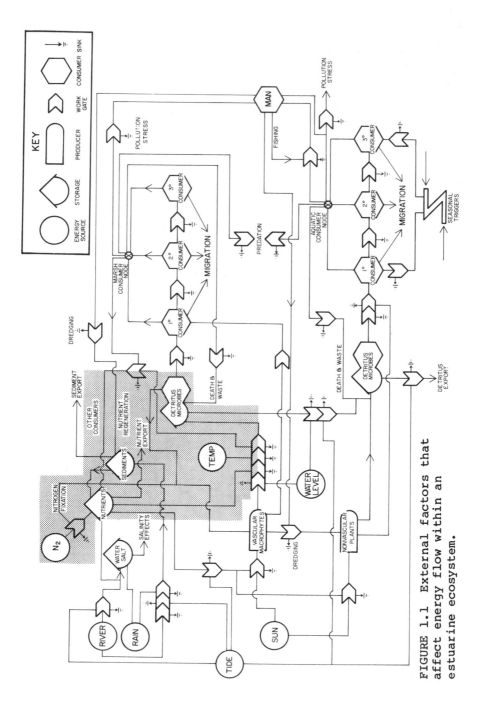

FIGURE 1.1 External factors that affect energy flow within an estuarine ecosystem.

the energy flow within an estuarine ecosystem, such as the
sun, tide, weather, runoff, pollution, fishing, and man
himself.[5]

The above illustration of energy flow, however,
represents only an essentially instantaneous image of
interactions. Large-scale and long-term dynamic changes
also occur. River flow commonly varies several orders of
magnitude within the average year and far more with exceptional events, such as floods. Large ranges in temperature can occur from summer to winter. Tides may be
semidiurnal, diurnal, or mixed. Tidal currents move vast
quantities of water through the system, changing the depth
and flooding and draining the edges of the estuary. These
tides vary both cyclically and irregularly; fortnightly,
spring to neap, over lunar periods and longer cycles; and
in close response to meteorological events, especially
storms. Even the mean sea level can vary on both short-
and long-time scales because of eustatic or local geologic
changes or changes induced by humans.

The hydrographic pattern in estuaries changes con-
tinually. Circulation is principally dependent on the
morphology of the estuary, riverine input, winds, and
differences in water density between estuary and ocean
and between parts of the estuary. All of these vary with
time. The classic pattern exists in many estuaries in
which the density structure causes net downstream transport of surface water, net upstream movement of saline
deep waters, and substantial mixing between. In each
estuary, this pattern is modified by many factors,
including tidal, seasonal, and meteorological effects.

Similar illustrations of complexity could also be
drawn from the transport and behavior of the dissolved
and particulate materials that perpetually enter these
systems, from the chemical processes that involve such
materials as nutrients or toxic substances, and from the
effects of both small-scale and large-scale engineering
activities.

Estuaries are remarkably resilient to natural events,
such as storms, seasonal temperature extremes, and changing freshwater flow; however, they are highly vulnerable
to the following kinds of anthropogenic changes:

1. Excessive burdens of natural materials (sediments,
nutrients, other normal chemicals) can overwhelm the ca-
pacities of estuaries for accommodation. An estuary may
recover if the excess of nutrients, for instance, is removed, but high sediment loading as a result of increased

human activity speeds the natural filling process and creates irreversible loss.[6] Through other processes, large-scale engineering changes for special purposes can modify the morphology of an estuarine system, redirect important volumes of freshwater to other systems, seriously modify the seasonal patterns of freshwater input, or alter the shoreline and important exchanges between land and bay.

    2. Estuaries are fragile under effects of powerful biocides that have only recently been invented. These biocides are alien to the genetic history of the estuarine organisms involved, which may have no mechanisms for defense. The principal food web in estuaries comprises a relatively small number of species; and an unseen wipe-out of a species could cause fundamental changes and diminish the utility of the system.

INTERDISCIPLINARY PROBLEMS--EXAMPLES

Many of the most critical needs for knowledge about estuaries can be met only through interdisciplinary fundamental studies. The following are important examples:

- In the San Francisco Bay system,[6] the proposed diversion of increasing quantities of water to southern California would affect (1) hydrographic behavior of the lower San Joaquin and Sacramento Rivers and of major portions of San Francisco Bay; (2) input of nutrients and other chemicals; (3) salinity distributions; (4) standing stocks of many biological species and communities; (5) distribution of predator and prey species; (6) input, movements, and behavior of sediments; and (7) water quality for many uses.

- In a study area of 11,500 square miles of coastal Louisiana, the highly productive marsh and wetlands are declining at the rate of approximately 40 sq. mi. per year.[7] In comparison, there was accretion in the nineteenth century and annual loss at rates of 6.7 sq. mi. in 1913, 15.8 sq. mi. in 1948, and 28.1 sq. mi. in 1967. The causes and effects involve the changes that have been made in the distributary outlets from the Mississippi River by engineering of levees, allocation of flow between the Mississippi and Atchafayla systems, decay of old deltas, hydrographic patterns, waves, storms, deposition rates, erosion rates, subsidence, modification of wetlands for petroleum exploration, flood-control practices, dredging, the production and nourishment of a variety of fish and shellfish important for human use, and other components and processes.[8]

● The development of large volumes of anoxic or nearly anoxic water off New Jersey in recent years, with its extensive threat to fish and shellfish and to the waste-assimilation capacity of the region, is apparently linked to meteorological sequences as well as to nutrient inputs from the adjoining New York metropolitan area.[9]

● In the Chesapeake Bay, the deep waters of the mid and lower Bay are progressively depleted of oxygen during mid and late summer in some years.[10] Depletion is not fully understood, but the causes involve (1) high inflow of freshwater during spring and warm weather, which increases stratification and reduces vertical mixing; (2) respiration in the deep water, which may be enhanced by organic debris from the previous year and which in turn may be increasing because of greater nutrient loading from the tributaries; and (3) perhaps benthic respiration. The effects of the development and extension of the depleted mass may be of biological importance, because such deep waters are potentially useful for fish, crabs, and other species and because the up-estuary transport of juvenile blue crabs, some juvenile fish, and other species occurs only in these deeper waters--although not necessarily at this season. Seiche-like rocking of the low-oxygen mass is suspected as the direct cause of substantial mortalities among fish, benthic species, and trapped blue crabs. Other biological relationships are modified. Chemically, anoxic conditions substantially alter the processes in the sediments and in the water column. Only well-designed interdisciplinary investigations can adequately link the physical, chemical, and biological components and provide predictive capability.

● Most efforts to understand the populations of estuarine species--efforts that are important not only for scientific objectives but also for the management of fisheries--are frustrated by our inability to deal simultaneously with the processes of physical transport (nutrients; planktonic food; eggs, larvae, and juveniles); biological relationships with prey, predators, and diseases; and the potentials for chemical injury or enhancement from toxic wastes or nutrients, as well as the effects of the migratory movements guided by estuarine conditions.

THE PATTERN OF PREVIOUS RESEARCH AND PUBLICATION

Much research has been devoted to estuaries. A review of
the journals and volumes makes clear the abundance of
disciplinary publications and the paucity of comprehensive
approaches that simultaneously investigate the interactive, physical, geologic, chemical, and biological components and processes in an effort to develop predictive
capability in the estuarine system.

Officer et al.[11] located 1699 refereed published
papers from a five-year period as gleaned from a wide
variety of scientific, engineering, and technical journals
and edited volumes. For the Chesapeake Bay, more than
9000 published reports exist, of which many, but not all,
were refereed prior to publication. Officer et al. also
found papers in 34 diverse journals and 19 edited volumes.
One American journal, *Estuaries*, and the international
journal, *Estuarine Coastal and Shelf Science*, are exclu-
sively devoted to estuarine and coastal papers. In addition, there is a large but unmeasured production of other
kinds of publications, especially in-house papers and reports. Many of these are known to be of high quality and
value, but they may not have been subjected to the rigors
of independent peer review.

Officer et al. also analyzed the institutional affil-
iation of senior authors of refereed papers and the iden-
tifiable federal funding of estuarine research. They
report that the academic community produced 77 percent of
the refereed publications, followed by federal laboratories (15 percent) and state, municipal, industrial, and
other sources (8 percent). An analysis of federal funding
related to ocean pollution research, development, and
monitoring reports that the projects in fiscal year 1978
were funded at a total of $164 million, of which $131
million was ascribed to research, including $40.6 million
related to estuaries.[12] Of that, the academic community
was granted $14.0 million or 37 percent of the estuarine
research funds.

Most of the estuarine research has been along disciplinary lines, devoted to relatively specific aspects of
biological, geologic, chemical, or physical processes. A
recent analysis by W. J. Hargis and T. M. Armitage (Virginia Institute of Marine Science, personal communication)
of published papers directly relating to the Chesapeake
Bay from 1971 to 1982 cataloged 2056 research products.
Specific disciplinary studies (1688) made up 82 percent
(946 in biology; 559 in physics or chemistry; 98 in eco-

nomics; and 85 in social, behavioral, or institutional studies). Many of these involved more than one narrow discipline, but only 18 percent were identified as inter-disciplinary in design or interpretation. Chesapeake research was heavily influenced by the program of Research Applied to National Needs (RANN) in the National Science Foundation, which emphasized an interdisciplinary ap-proach, but the program has been terminated.

There has been notable progress in the understanding of estuaries during recent decades. The emergence of the two journals devoted to coastal and estuarine papers, both of which have backlogs of good manuscripts, is one evidence of production of knowledge. A new *Journal of Shellfish Research* was initiated in 1981 and will be dominated by coastal studies. The Estuarine Research Federation, a confederation of societies representing New England, the Mid-Atlantic, South Atlantic, Gulf of Mexico region, and Pacific Coast, was established in 1971; it has conducted biennial international conferences that have produced a series of volumes.[13] Other edited volumes entirely or substantially directed to research on estu-aries and the inshore coastal region are listed by Officer et al.[11] and in note 14 at the end of this chapter.

RECOMMENDATIONS

The limits to success in balancing the uses of an estua-rine system for any set of desired objectives are set by the available fundamental understanding of the components and processes of the system. Empirical trial-and-error management and decision-making techniques have been em-ployed in matters pertaining to estuaries, but these techniques involve high risks because of the value and potential fragility of estuaries. The store of reliable knowledge and predictive capability is the primary source of useful guidance when decisions must be made.

Because of the complexity of the components and pro-cesses in estuarine systems and the need for this knowl-edge in managing these systems, the Panel recommends the following:

1. *The primary focus of future research in estuaries should be on interdisciplinary relationships.* However, there needs to be an explicit recognition that disciplin-ary investigations provide the building blocks necessary to an interdisciplinary framework. Although the disci-

plinary knowledge bases are currently incomplete, they
are adequate to start establishing the interdisciplinary
understanding.

Productive research on such complex problems is now
feasible. Much is known of the various components and
processes, and even of the interactions. Research facilities, including well-equipped vessels, can be simultaneously brought to focus on appropriate facets of the
problem. Computer and analytic capabilities have advanced
to a point that will permit a combination of theoretical
and mathematical efforts that were previously impossible.
Scientific experience and expertise now available in systems modeling and analysis can also be attracted to this
kind of challenging problem.

However, it is exceptionally difficult to design and
conduct such complex studies--and even more difficult to
obtain adequate funding for an appropriate period. The
Panel is satisfied that the scientific community will
respond with competence and exceptional achievement if
funding becomes available. Team study, the use of con-
sortia, *ad hoc* linkages, and other cohesive procedures
have frequently been effective in other areas of earth
science (particularly if coordination is an integral
element in the design of such studies), and they can be
developed and used for estuarine studies.

Examples of possible research focuses developed within
the remainder of this report are the following:

(a) Unless the estuary in question is studied
carefully and thoroughly, some of the most important processes that affect it may go unnoticed; however, it is
unreasonable to consider, in detail, each and every relationship between all the species of an estuary. This
points to the necessity of conducting detailed spatial and
temporal surveys of estuaries for qualitative and quantitative stock assessment of plant and animal populations
and combining this assessment with the nature of the
variations in circulation and chemistry.

(b) Once the more important and quantitatively
significant trophic and other interactions between species
of an estuary are understood, research should consider how
environmental stress of various kinds, natural and anthropogenic, affect those interactions and the general structure and function of the ecosystem. This may be done
empirically, by observing changes in community organization in time or space following the onset of a given
stress (e.g., a new waste-water discharge, a freeze, or a
drought). The cause and effect of such changes may be

obscure even when the interactions of the community components are reasonably well known.

(c)   The determination of Lagrangian (moving coordinate system) particle trajectories within estuaries exhibiting complex Eulerian (fixed coordinates) flow patterns is fundamental to a description of the movement and dispersion of materials within an estuary and their ultimate exchange to the adjacent water. A fundamental problem that exists is how best to develop our ability to provide a Lagrangian description of flow within estuaries and whether this is, of necessity, dependent on numerical simulations.

(d)   Mechanisms for the generation of nontidal circulation patterns through residual vorticity produced both by the interaction of the tidal stream and surface wind stress with the bathymetry should be investigated. This problem is amenable to analytical, numerical, and experimental investigation. Useful experiments designed to investigate residual vorticity generation could be accomplished using moored current-meter arrays in areas of complex bathymetry.

(e)   Mechanisms responsible for the often rather abrupt transition from stratified to vertically well-mixed conditions should be investigated, as well as the changes in internal circulation that accompanies this transition. The nature of these transitions needs to be investigated in relation to the suspended-sediment load and chemical distribution, particularly as these transitions affect the distribution of nutrients and the quality and quantity of light vital to the biological activity.

(f)   Research is needed on how sediments are re-distributed as the results of tidal currents and episodic storms. What is the net effect in the coastal zone of estuarine flow and wind-driven upwelling or downwelling on the accumulation or exportation of sediments and their subsequent effects on chemical and biological distributions?

(g)   The role of bioturbation in releasing chemical species to the water, in increasing the trapping efficiency of the sediment for other chemical species, and in the redistribution of nutrients needs to be investigated.

(h)   The effects of episodic storms and floods on the biota, sediment load, and chemistry of estuaries need further investigation. A necessary component of assessing these effects is the existence of baseline survey data before such an episodic event in order to assess the change and in determining the time needed and processes necessary to restore the estuary to the pre-episodic event.

2. *Such interdisciplinary research in both university and governmental laboratories should receive increased support to provide the basic framework for informed management of estuarine systems.* Research useful to the management of estuarine systems can take two closely related approaches. Baseline studies that determine the state of the system are necessary to assess the effects of both natural and anthropogenic change. The other approach involves understanding estuarine processes in order to develop and verify predictive modeling (what will be the result if part of an estuarine system is changed). The predictive modeling utilizes the baseline studies as a primary data base. The country's economic interests in estuarine resources are too great to allow us to continue with a trial-and-error management approach.

NOTES AND REFERENCES

1. N. C. Matalas, J. Maciunas Landwehr, and M. G. Wolman (1982). Prediction in water management, in *Scientific Basis of Water-Resource Management*, Geophysics Study Committee, National Research Council, National Academy Press, Washington, D.C., pp. 118-127.
2. National Marine Fisheries Service (1981). *Fisheries of the United States, 1980*, National Oceanic and Atmospheric Administration/Department of Commerce, Washington, D.C., 96 pp.
3. B. H. Ketchum, ed. (1972). *The Waters Edge--Critical Problems of the Coastal Zone*, MIT Press, Cambridge, Mass., 393 pp.; NRC Environmental Studies Board (1973). *Water Quality Criteria 1972*, National Academy of Sciences, Washington, D.C., 594 pp.; U.S. Environmental Protection Agency (1977). *Estuarine Pollution Control and Assessment, Proceedings of a Conference*, U.S. Government Printing Office, Washington, D.C., 755 pp.
4. NRC Environmental Studies Board (1981). *Testing for the Effects of Chemicals on Ecosystems*, National Research Council, National Academy Press, Washington, D.C., 98 pp.

15

5. National Coastal Ecosystems Team (1980). *Ecological Characteristics of the Sea Island Coastal Region of South Carolina and Georgia--Resource Atlas*, Fish and Wildlife Service, U.S. Department of the Interior, FWS/OBS-79/48, 56 pp.
6. T. J. Conomos, ed. (1979). *San Francisco Bay: The Urbanized Estuary*, American Association for the Advancement of Science, Washington, D.C., 493 pp.
7. S. M. Gagliano, ed. (1976). *Special Report on Marsh Deterioration and Land Loss in the Deltaic Plain of Coastal Louisiana*, Coastal Environment, Inc., Baton Rouge, Louisiana, 7 pp.
8. R. H. Chabrack, ed. (1973). *Proceedings of the Coastal Marsh and Estuary Management Symposium*, Louisiana State University, Baton Rouge, Louisiana, 316 pp.
9. J. H. Sharp, ed. (1976). *Anoxia in the Middle Atlantic Shelf during the Summer of 1976*, National Science Foundation, Washington, D.C., 122 pp.
10. J. L. Taft, W. R. Taylor, E. O. Hartwig, and R. Loftus (1980). Seasonal oxygen depletion in Chesapeake Bay, *Estuaries 3*, 242-247.
11. C. B. Officer, L. E. Cronin, R. B. Biggs, and J. R. Ryther (1981). A perspective on estuarine and coastal research funding, *Environ. Sci. Technol. 15*, 1282-1285.
12. Interagency Committee on Ocean pollution Research, Development, and Monitoring/Federal Council for Science, Engineering, and Technology (1979). *Federal Plan for Ocean Pollution Research, Development, and Monitoring, Fiscal Years 1979-1983*, 160 pp.; Interagency Committee on Ocean Pollution Research, Development, and Monitoring/Federal Council for Science, Engineering, and Technology (1979). *Federal Plan for Ocean Pollution Research, Development, and Monitoring, Fiscal Years 1979-1983*, Working Paper 1, 230 pp.; Interagency Committee on Ocean Pollution Research, Development, and Monitoring/Federal Council for Science, Engineering, and Technology (1979). *Reports of the Subcommittee on: National Needs and Problems; Data Collection; Storage and Distribution; Monitoring; Research and Development*, Working Papers 0-5, 177 pp.
13. L. E. Cronin, ed. (1975). *Estuarine Research*, Estuarine Research Federation, Academic Press, New York, 587 pp.; V. S. Kennedy, ed. (1980). *Estuarine Perspectives*, Estuarine Research Federation, Academic Press, New York, 533 pp.; V. S. Kennedy, ed. (in

press). *Estuarine Comparisons*, Estuarine Research
Federation, Academic Press, New York; M. L. Wiley,
ed. (1977). *Estuarine Processes*, Estuarine Research
Federation, Academic Press, New York, 428 pp.; M. L.
Wiley, ed. (1978). *Estuarine Interactions*, Estuarine
Research Federation, Academic Press, New York, 603 pp.

14. T. W. Dake, ed. (in press). *Man's Impact on the
Coastal Environment*, U.S. Environmental Protection
Agency; E. D. Goldberg, ed. (1979). *Proceedings of
a Workshop on Scientific Problems Relating to Ocean
Pollution*, Estes Park, Colo., July 10-14, 1978, U.S.
Dept. of Commerce, National Oceanic and Atmospheric
Administration, Environmental Research Laboratories,
225 pp.; E. D. Goldberg, ed. (1979). *Proceedings of
a Workshop on Assimilative Capacity of U.S. Coastal
Waters for Pollutants*, Crystal Mountain, Wash., July
29-August 4, 1979, U.S. Dept. of Commerce, National
Oceanic and Atmospheric Administration, Environmental
Research Laboratories, 284 pp.; B. Kinsman, J. R.
Schubel, M. J. Bowman, H. H. Carter, A. Okubo, D. W.
Pritchard, and R. E. Wilson (1977). *Transport Process
in Estuaries: Recommendations for Research*, Special
Report 6, Marine Science Research Center, State Uni-
versity of New York, Stony Brook, N.Y., 21 pp.; B. J.
Neilson, and L. E. Cronin, eds. (1981). *Estuaries and
Nutrients*, Chesapeake Research Consortium, Inc., Hu-
mana Press, Clifton, N.J., 643 pp.; NRC Ocean Affairs
Board (1971). *Marine Environmental Quality--Suggested
Research Programs for Understanding Man's Effect on
the Ocean*, National Academy of Sciences, Washington,
D.C., 187 pp.; NRC Committee on Oceanography and NRC
Committee on Ocean Engineering (1971). *Waste Manage-
ment Concepts for the Coastal Zone--Requirements for
Research and Investigation*, National Academy of
Sciences, Washington, D.C., 126 pp.; NRC Geophysics
Study Committee (1977). *Estuaries, Geophysics, and
the Environment*, National Academy of Sciences, Wash-
ington, D.C., 127 pp.; The Johns Hopkins University,
University of Maryland, and Virginia Institute of
Marine Science (1971). *The Chesapeake Bay, Report
of a Research Planning Study*, Report to the National
Science Foundation, Washington, D.C., 211 pp.

2   ENVIRONMENTAL EFFECTS
ON ESTUARINE BIOTA*

The biota is important to the fundamental processes in
estuarine systems and to almost all human uses of these
systems, especially food production, aesthetic enjoyment,
and the assimilation of wastes. The most valuable and
useful question to pose when physical, chemical, or
biological alterations are considered is usually: What
effect will this have on the biota? Predictive estimation
of such effects have been explored, but the following
discussion of efforts and needs demonstrates that the
requisite research will be complex. It must involve several levels of biota and ecosystem components and processes and cannot be fully successful without interdisciplinary linkage to the physical and chemical components and
changes in an estuarine system. Attention in this chapter
is given to the responses of individuals, species, and
communities. The roles and challenges of system modeling
are discussed, and the importance and problems of usefully
documenting alterations in the estuarine ecosystem and of
estimating the integrated condition of such systems are
presented as research challenges.

*This chapter was largely developed by a subpanel on
estuarine biota, which consisted of the following: John
H. Ryther, Harbor Branch Institution, *Subpanel Leader*;
L. Eugene Cronin, Chesapeake Research Consortium; Robert
Feller, University of South Carolina; Donald Heinle,
CH_2M-Hill; A. Fred Holland, Martin-Marietta Laborato-
ries; and Theodore Smayda, University of Rhode Island.

EFFECTS ON INDIVIDUAL ORGANISMS

Studies that investigate the effects of various environmental stresses on the survival, growth, or other vital
functions of individual aquatic species are among the
oldest forms of ecological research. Such studies were
intended to provide information on where and when organisms could occur in nature and what factors of the
environment limit and control their abundance and distribution; this approach is now recognized as being too simplistic in itself for general application. These studies,
however, may provide useful information for range prediction, control, or management purposes in situations where
certain species suddenly become a dominant or key component of an estuarine ecosystem.

A similar type of autoecological laboratory study has
recently been pursued, particularly by the U.S. Environmental Protection Agency and its contractors, in which the
effects of anthropogenic stress (i.e., toxic pollutants)
on aquatic organisms are determined. These studies are,
for the most part, designed to determine lethal concentration thresholds of toxic contaminants for the purpose
of establishing discharge guidelines or standards (NRC
Environmental Studies Board, 1981; Cairns, 1982). Bioassay studies of this nature are not, as such, particularly useful research tools for understanding estuarine
processes.

Assuming that an ultimate objective is to understand
and to be able to predict the complex dynamics of the
estuarine ecosystem as a unit, reliable data need to be
acquired on the rates of such fundamental processes as
feeding, food conversion efficiency, metabolism, growth,
maturation, fecundity, and reproductive success of as many
as possible of the key organisms of a given estuary. Rate
data on these processes must be determined as functions of
the environmental variables that have significant effects,
of which temperature is an obvious example. This type of
information is, unfortunately, still rare except for a few
species of plankton and some commercially important finfishes and invertebrates (the latter frequently as a spin-
off from aquaculture research).

Detailed physiological rate studies of all the organisms that live in estuaries or even one representative
estuary is not reasonable. Attention should be focused
on numerically or functionally dominant (or key) species,
with emphasis on species that have usually been neglected
in the past and about which too little is now known to

attempt even rough approximations of the physiological
rates. These include, for example, some of the small
estuarine organisms: the meiobenthos, the microzooplankton
and protozoa, and the bacteria including the autotrophic
cyanobacteria of newly discovered ecological significance.
Another group of organisms that periodically dominate many
estuarine ecosystems and on which there is little information are the soft-bodied macrozooplankton, the jelly-
fishes, tunicates, and ctenophores. Another important but
often neglected group of organisms is that of the smaller,
commercially unimportant finfishes that often outnumber
and may outrank in functional significance the more
familiar food species.

Simple feeding, growth, and metabolic rate measurements, perhaps as a function of temperature, of a small,
selected group of the above kinds of organisms would
provide a key element for the long-range objective of
predictive numerical estuarine-ecosystem modeling.

The autoecological laboratory studies described above
must, however, be proceeded or accompanied by appropriate
field studies, without which neither the most important
organisms nor the appropriate physiological process to
study can be determined. Before feeding rates are measured, for example, it is necessary to know what food or
foods the animal in question utilizes in the estuary, a
characteristic that is often poorly known even for some
of the most thoroughly studied species. It is here that
autoecology and synecology merge and that laboratory and
field studies become mutually interdependent and comple-
mentary.

EFFECTS ON COMMUNITIES

To assume that one can study the effects of environmental
stress on estuarine communities and ecosystems is to imply
that the normal structure and function of estuarine ecosystems is already well understood. Such, unfortunately,
is far from the truth. Ecosystem components may, in some
cases, be described in qualitative and even in numerical
terms by means of exhaustive field surveys, but even the
most painstaking of surveys fail to reveal the intricacies
of the relationships, trophic and other, between its
members.

Estuaries are, in fact, more complex in this respect
than other aquatic ecosystems, with both freshwater and
marine inputs and usually at a shallow depth where the

pelagic and benthic communities functionally merge into one. Jellyfish in Chesapeake Bay have been observed to settle and feed on the bottom during parts of the day. The adult hard clam, *Mercenaria*, living on the bottom of Narragansett Bay has been estimated to equal or even exceed zooplankton as a grazing influence on phytoplankton abundance and species composition. Meroplankton, the temporary larval forms of larger benthic or pelagic invertebrates, may far exceed copepods and other holo- plankton organisms in biomass and in feeding intensity during the brief periods when they are present, and each group (fish, crustacea, echinoderms, molluscs) may have a different annual periodicity as well as different feeding habits, food preferences, and other behavior. Phytoplank- ton production is much more closely coupled with the ben- thos in estuaries than is the case in the ocean. Algal blooms, which are not immediately consumed by zooplankton because of the lack of synchrony between the plant and animal populations, can quickly sink to the bottom before the cells have time to decompose and recycle their nu- trient contents. As much as 90 percent of a spring diatom bloom may end up on the bottom of an estuary, in contrast to 10 percent or less in the ocean, resulting in benthic communities orders of magnitude more dense in estuaries. The bottoms of shallow estuaries may also support their own producer communities in the form of both unicellular and macroscopic algae, as well as higher rooted vegeta- tion, and the productivity and ecological role of this benthic flora are not well understood.

Another factor that adds to the complexity of estua- rine communities is that of migratory populations that may enter and leave the estuary or its tributaries, seasonally or aperiodically, to spawn or feed, actively or passively, and predictably or unpredictably. Large populations of small bottom-feeding finfishes, such as spot and croaker, move into the tributaries of Chesapeake Bay where they may devastate the benthos before they leave--a significant factor affecting the flux of organic matter and nutrients into and out of these areas that may be missed completely if their movement is not observed during the few days of the year in which it occurs. Similarly the soft-bodied jellies (coelenterates, ctenophores, tunicates) may move into estuaries virtually overnight and decimate zooplank- ton communities with disastrous consequences to larval and juvenile resident planktivorous finfish populations, yet their impact has often been ignored completely in simplistic trophic-dynamic models.

A major direction of modern fisheries biology is that of multispecies management, a concept that recognizes the trophic interactions of the several commercial species with themselves and/or a common food supply. Heavy exploitation or poor annual recruitment of a planktivorous species such as herring in a coastal fishery may result in an increased yield of benthic fishes as the uneaten food sinks to the bottom. No such simple relationship occurs in estuaries, where, as noted above, the planktonic and benthic biota are one community and where year-class success of a food species is as apt to result from competition for space as for food. Fishing pressure may exert an influence on the abundance or species composition of the food supply of an estuary, even by the offshore capture of migrant fishes hundreds of miles away.

The foregoing examples are intended to illustrate the extreme complexity of the estuarine community. Many more could be given. It is the first job of the ecologist to understand these interactions qualitatively and to try to assess their relative importance and magnitude in the normal flow of matter and energy through the ecosystem. To consider in detail each and every relationship between all the species of an estuary is unreasonable, but as is evident from the previous discussion, unless the estuary in question is studied carefully and thoroughly, some of the most important processes that affect it may go unnoticed. This fact points to the necessity of conducting detailed spatial and temporal surveys of estuaries for qualitative and quantitative stock assessment of plant and animal populations.

However, the abundance and distribution of species still does not reveal many of the subtleties of their interactions, for which there is often no substitute for direct observation. There is still a need in estuarine research for the old-fashioned natural historian whose efficiency may be immeasurably increased today by means of modern life support, television, photography, electronic data recording, and other submarine equipment and technology.

Once the more important and quantitatively significant trophic and other interactions between species of an estuary are understood, one may begin to consider how environmental stress of various kinds, natural and anthropogenic, affect those interactions and the general structure and function of the ecosystem. This may be done empirically, by observing changes in community organization in time or space following the onset of a given stress (e.g., a new

waste-water discharge, a freeze, a drought). The cause
and effect of such changes may be obscure even when the
interactions of the community components are reasonably
well known. For example, various trace contaminants
(hydrocarbons, heavy metals) through selective toxicity
tend to cause a shift in phytoplankton community structure
from large diatoms to small flagellates. Such a change
is often considered undesirable. Diatoms are the first
step in a food web that normally leads to copepods and
finfishes, whereas a flagellate-dominated phytoplankton
community often supports a grazing community of soft-
bodied invertebrates that represent a trophic dead end.
However, this type of shift is not consistent. In the
Marine Ecosystem Research Laboratory (MERL) experiments
at the University of Rhode Island's Narragansett Marine
Laboratory, ecosystems experimentally stressed with oil
resulted in a bloom of diatoms. In that case, the phytoplankton shift resulted from the toxicity of the added
hydrocarbon to benthic animals that had exerted grazing
pressure on and supressed the diatoms. The latter kind
of interactions is difficult to detect and virtually impossible to prove through simple observation of community
change in a natural environment. One sees the end result
and, if appropriate baseline studies have been conducted,
may determine that a community alteration has occurred,
but the mechanisms that caused it remain obscure.

Probably the only way in which the actual interacting
community processes resulting from environmental stress
can be identified and documented is by means of experimental ecosystems. These may be simple or complex, large
or small, and of a wide variety of shapes and configur-
ations. The large and rather complicated systems such as
MERL (12 cylindrical tanks, about 2 m in diameter and
5.5 m high, of 13 m^3 volume) and the National Science
Foundation-International Decade of Ocean Exploration-
sponsored Controlled Ecosystem Pollution Experiment
(CEPEX) (three units, 9.5 m in diameter and 29 m deep, of
1700 m^3 volume, that were submerged in Sannich Inlet,
Vancouver, B.C.) have both been rather troublesome to
operate over the long periods of time necessary for community equilibria to become established. The high cost
of the CEPEX units also made it difficult to achieve adequate replication and experimental controls. Smaller experimental tanks, 1 m^3 or less in volume, even including
plastic trash barrels, often suffice for certain kinds of
experiments and may be replicated many times at modest
price.

It is usually preferable to conduct several experiments including different levels of an environmental stress or even several stresses, suitably replicated and with adequate experimental controls, all simultaneously rather than in sequence. A typical ecosystem experiment can last for periods of one to several months, and sequential experiments will therefore necessarily occur over two or more seasons during which environmental conditions other than those being tested are certain to vary significantly. The alternative of conducting such studies in controlled environmental chambers is so costly as to be usually prohibitive, and the chances of mechanical failure of some part of the system over prolonged periods of time are so high as to be almost certain.

Not only do small, simple experimental ecosystems lend themselves to the possibility of conducting several replicated experiments simultaneously, but the simpler the experimental design, the less likely that experiments will be aborted because of mechanical failure. There are, of course, limitations to the kinds of experiments that can be conducted and, in particular, the complexity of the ecosystem that can be studied in small, simple systems. Accordingly, there is a need for a range of such experimental systems, both in size and complexity.

Even large experimental ecosystems carefully inoculated with all the significant constituents of a natural estuarine environment cannot and will not duplicate events in the natural system. Rather, it must be considered as a microcosm unto itself that will respond to the interactions of its organisms and to the environmental stresses to which it is exposed. These reactions will be understandable and interpretable only by comparing them with similar experimental systems that differ to some slight degree—usually one species or unit of environmental stress—not by comparing them with the natural environment.

It follows that the simpler the experimental system and the more accurately it can be replicated and controlled, the greater will be the chance of observing reproducible results. At the same time, the more simple the experiment and the more removed it is from nature, the interpretation of results in terms of natural events becomes more difficult and tenuous.

Another approach to experimental estuarine ecosystems might involve the creation of artificial estuaries by digging canals or channels off a natural estuary or restricting a small, manageable arm of an estuary by netting

or some other barrier that would permit the natural flow
of water but prevent the movement of most organisms. It
is even possible to replicate such experimental estuaries
so that the effects of stress added to one may be compared
with an unstressed control. To our knowledge, few if any
attempts have been made to do this, but the concept would
appear to have some advantages over purely artificial
systems (e.g., tanks, plastic bags) in that a more natural
and more complete community would be available for experimental manipulation, and it is quite possible that unan-
ticipated community interactions could result in alterations quite different from those obtained from artificial
systems.

Finally, experiments may be conducted within estuaries
on completely natural communities, for example, excluding
benthic or intertidal organisms from predators by con-
structing wire-mesh cages over them, enclosing parcels of
the bottom or water and stressing them with toxic pollutants, and introducing predators or competitors to an
estuarine area in which they do not naturally occur.

All the above experimental methods are useful in
providing insight to the mechanisms by which environmental
stresses alter estuarine ecosystems. They do not, for the
most part, provide the kind of quantitative data that may
be used for predictive or regulatory purposes. But
through the combination of identifying the significant
trophic interactions of estuarine ecosystems through field
observations and experimental ecosystem studies of various
kinds and the experimental measurement of the rates of
these identified processes, it may ultimately be possible
to express the effects of stress on the estuary as an
ecological unit in quantitative terms through appropriate
numerical modeling techniques.

ECOSYSTEM MODELING

While predictive numerical modeling of complete ecosystems
should be considered as the long-range goal of ecological
research, it should be emphasized that it is not realistic
for the immediate or near future, particularly in the
complex estuarine environment. As pointed out earlier,
it is first necessary to obtain basic information on the
nature and rates of at least the more important trophic
and other interactions of the key organisms as well as
stock sizes and distributions. This does not mean, however, that there is not a place for the simpler heuristic
type of models in contemporary estuarine research.

Simple, empirical predictive procedures, often referred to as models, have been and are being used for estimating or forecasting various aspects or functions of the plankton, fish, and shellfish populations of estuaries. As an example, the timing of the onset of spawning or the setting of spat of oysters (important for the placing of culch or other devices for collecting the oyster spat and avoiding other epibiota) can often be correlated with temperature and/or salinity. The species composition, density, and distribution of phytoplankton blooms in Great South Bay and Moriches Bay, Long Island, have been related to the source and nature of their nutrient source (duck farms located on tributaries to the bays), temperature, salinity, wind speed and direction, and the flushing rate of the bays. The year-class success of striped bass in Chesapeake Bay, currently in serious decline and believed by many to have suffered the impact of antropogenic stresses, is now believed to be dependent on a combination of climatic factors (temperature, rainfall) in the upper regions of the estuary during the critical postspawning period. Elaborate models have been developed to predict the loss of striped bass larvae through entrainment in power-plant cooling water systems in the Hudson River, and many similar predictions have been made involving other species and other utilities, existing or planned. A reasonable short-term goal of estuarine ecosystem studies is to try to understand why simple predictive procedures do or do not work, to refine and improve on them, and to arrive at other gross predictive procedures for those fortuitous cases where only a few environmental factors control important biological events.

Somewhat more sophisticated numerical modeling can and should be employed as an ancillary tool to help guide and direct basic studies of estuarine ecosystem structure and function. Such modeling may be used to predict what *might* happen if certain processes were the controlling factors. Modeling can point out what components, rate constants, processes, and interactions are both poorly known and of sufficient potential significance to merit special attention. Modeling may also point out what relative effects specific processes may have on other parts of the ecosystem.

For example, for a model of a relatively simple aquatic ecosystem consisting of nutrients, phytoplankton, and zooplankton, the numerical relations are intricate. They involve a set of temporally and spatially dependent, coupled differential equations involving one or more nutrient conservation equations, one or more phytoplankton community equations, and one or more zooplankton community equations. Each equation includes both the hydrodynamic advective and nonadvective terms and the constituent source and sink, growth and decay, and transformation and coupling terms.

In some cases, particularly those for lakes and the deep ocean where the hydrodynamics affecting the ecosystem can be defined in somewhat simpler terms than that for estuaries, analytical solutions may be given. In other cases, scientific modeling solutions, though limited in their applicability, can be derived that do point out the potential importance of various factors such as nutrient limitation, zooplankton grazing, self-shading, hydrodynamic residence time, and mixing effects on phytoplankton populations.

DOCUMENTATION OF ECOSYSTEM ALTERATION

Estuarine ecosystems are often assumed to have changed in historical times and are, in fact, now in the process of alteration. The implicit changes are not often well defined but generally have to do with such human-use values such as aesthetic or recreational enjoyment or commercial food production. Most often, the change is also believed to have occurred in the direction of a deterioration of such values. Unfortunately, there are not often many scientific data to substantiate or document such changes in specific estuaries, not to mention the associated problem of trying to relate such changes, if they have occurred, to some one or more causes, anthropogenic or other. If for no other reason, current thorough descriptive accounts of representative estuarine ecosystems are needed to ensure that future alteration may be well documented.

Examples of the kind of study that can provide useful baseline data against which future change may be assessed is the recently completed *Environmental Atlas of the Potomac Estuary* prepared for the state of Maryland (Power Plant Siting Program) by Martin Marietta Corporation, the eight-volume *Comprehensive Study of San Francisco Bay* produced by the staff of the College of Engineering and School of Public Health, University of California at Berkeley, in the 1960's, and the ten-volume *Delaware Bay Report Series* published under a grant from the National Geographic Society. There are many other examples of estuarine baseline studies, particularly during the past decade, when state and federal regulatory agencies first began to require electric power utilities and other industries to make studies prior to the construction of facilities impinging on the estuarine environment. Designed as part of a "before and after" type of study, to demonstrate any measurable effect of the industrial operation in question after it went into operation, such studies have more recently lost favor to predictive modeling efforts as a more useful tool for decision making before the fact (the severe limitations of the modeling approach notwithstanding).

Baseline studies are useful only in direct proportion to their quality. Poorly conceived or executed surveys are not only wasteful of time and effort but may lead to erroneous conclusions. The San Francisco Bay study referred to above and others were closely scrutinized by Nichols (1973) with the following commentary:

> Efforts to describe the structure and function of benthic communities of the bay and to quantify the effects of waste discharge on them have been hampered by inconsistent and often faulty sampling methodology and species identification. Studies made show a lack of information on the normal life processes of the organisms concerned. The diversity index (a mathematical expression of the number of kinds of organisms present at a location), commonly used to describe the "health" of the benthic community, has been employed without regard for the need for standardizing methodology and species identifications or for understanding natural biological processes that affect such mathematical indices. There are few reliable quantitative data on the distribution of benthic organisms in the San Francisco Bay with which future assessments of the "health" of the benthic community might be compared. Methods for study of the benthos must be standardized, identifications of species verified by trained taxonomists, and new field and laboratory studies undertaken before we can expect to obtain an accurate description of the benthic fauna and its relations with the environment.

Commercial fish landings from larger estuaries represent some of the oldest and most complete relevant data that are available to document changes in the ecosystem. How well they relate to stock size of the species in question is problematical. In some cases, the correlation is thought to be high, as in the case of striped bass in the Chesapeake Bay. More often, social and economic factors affecting fisheries so confuse the issue that landings can be expected to bear little relationship to population size.

There are certainly a few cases, particularly with respect to the larger estuaries where there are one or more well-established marine laboratories located in the close vicinity, where old environmental data are available. Care must be taken in the interpretation and use of such information, as techniques and equipment for measurement and sample collection have changed significantly in the past decade or two. Those parameters that are simple and easiest to measure such as temperature, salinity, phosphate and silicate concentrations, and secchi-disc transparency are the most reliable. The ability to determine seawater nitrogen chemistry has greatly improved in recent years and now includes routine measurement of the concentration of ammonia, often urea, and sometimes total dissolved organic nitrogen, all of which are equally or more important than nitrate and nitrite--the fractions originally and not very accurately measured. Fluorometric determination of chlorophyll, as an index of phytoplankton biomass, is also a relatively recent measurement not comparable with older data obtained by other methods.

Similarly, comparison of old and new data on the species and numbers of the biota of different major groupings--for example, the phytoplankton, zooplankton, benthic invertebrates, fishes--may be difficult because of differences in the collection, sample processing, or preservation techniques used. Phytoplankton was routinely sampled with fine-mesh nets until it was discovered, some 20 or more years ago, that only the larger fraction of the unicellular algae--the 5-10 percent of the community greater than about 75 μm in diameter--were taken by even the finest nets. Subsequently, whole-water samples have been collected and preserved, the cells sedimented in settling chambers and counted by inverted microscopy. Today still another fraction of the autotrophic plankton community, the tiny cyanobacteria 1-2 μm in size that have usually gone completely unobserved by conventional microscopic examination, are found to be functionally a highly significant part of the photosynthetic or "producer" community. There has been a similar history in the earlier neglect and more recent inclusion of the smaller benthic invertebrates (meiofauna), the microzooplankton, and the smaller, noncommercial species of finfishes.

Despite these difficulties, there are certainly historical data sets, published and unpublished, that can provide useful information when interpreted with care on changes, permanent and reversible, cycles, and trends. The location, screening, processing, and synthesis of such data is a research undertaking in its own right, particularly where raw, unprocessed, and unpublished data collections are involved, the value, interpretation, and significance of which require value judgment by trained scientists. A notable example of environmental change in an estuary that has been carefully documented from the scientific literature and the historical record is the report *Historical Review of Water Quality and Climatic Data from Chesapeake Bay with Emphasis on Effects of Enrichment*, submitted to the Environmental Protection Agency by members of the Chesapeake Research Consortium, Inc. (Heinle *et al.*, 1980). Following is the draft summary from that manuscript:

Review of the available data on water quality in Chesapeake Bay has revealed changes caused by enrichment with nutrients. In the upper and middle Bay, and several tributaries, concentrations of algae present during the summer months have increased. There have been decreases in the clarity of the water associated with the increased algal stocks. Nutrient concentrations have also increased, phosphorus more notably so than nitrogen. In some of the tributaries, such as the Patuxent for which we have the best data, increased algal production has led to reduced concentrations of oxygen in the lower two-layered part of the estuary. The variations in concentration of oxygen are more extreme in surface waters than in the early 1960's in the Patuxent. Oxygen concentrations in the open Bay have not changed greatly, with the possible exception of extreme conditions during periods of extensive ice cover.

There have been historical variations in the abundance of commercial fishery stocks that are closely related to climatic variations. Beginning in 1969 or 1970, however, stocks of all anadromous spawning species and marine spawners representing higher trophic levels have declined to new long-time lows. The only exceptions are menhaden (marine-spawning

planktivorous fish) and bluefish (marine-spawning top
predators). That same time interval has, however,
been a period of above average rainfall and reduced
salinities in the Bay, making any conclusions concerning effects of enrichment difficult to achieve.
There is evidence for progression of the effects
of enrichment down the Bay with time. Increased inputs of nutrients may well lead to major changes in
the lower Bay, similar to those already observed in
the tributaries and middle Bay.

It is usually difficult or impossible to separate the
effects of natural environmental stresses from those
caused by man, an unanswered problem recognized by fishery
biologists. Weekly quantitative plankton samples and
related environmental data collected in Narragansett Bay
since 1959, first by D. M. Pratt and subsequently by T.
J. Smayda, both of the University of Rhode Island, have
revealed trends somewhat similar to those described above
for Chesapeake Bay. Over a 20-year period the maximum
populations characterizing the winter-spring phytoplankton
bloom have varied tenfold in density, as have summer maxima. But in 6 of the 9 years between 1970 and 1978, the
annual peak abundance of algae occurred in summer, in
contrast to the winter-spring annual maximum that occurred
every year prior to 1970. During that time, there was
also a doubling of both phytoplankton and zooplankton
biomass, an increase in primary production by 50 percent,
and a decline in stocks of winter flounder. All three
changes closely correlated. It is tempting to ascribe
them to human enrichment (eutrophication) of the Bay,
though this was not obvious from just nutrient loadings.
However, since 1978, there appears to have been the beginning of a return to the earlier conditions, with a winter-
spring annual phytoplankton maximum, a decline in plankton
biomass, and an increase in winter flounder stocks.
In both the Chesapeake Bay and Narragansett Bay data
there are suggestions of a 10-20-year cycle in the abundance of various populations, particularly of some of the
commercial finfishes and invertebrates, similar to and
closely following recognized climatic cycles. Robert Dow
of the Maine Department of Natural Resources pointed out
some time ago the same correlation between climatic cycles
and those of several commercial species in the estuaries
of that State (Dow, 1964). Presumably, if the apparent
relationship of stock size to climate is real, it should
be evident over a widespread region and testable by study-

ing the records from several estuaries over the same time periods, thereby perhaps separating anthropogenic from natural effects.

As far as we are aware, a science of "comparative estuarine ecology" has not yet become developed, but it would appear to be a promising field of research. Comparative studies of estuaries in urbanized areas with those in less-developed regions of comparable climatic and topographic characteristics could provide an insight to the long-term effects of anthropogenic stresses in the absence of (or as a supplement to) historical records. It would perhaps be necessary to travel far afield to find estuaries not heavily impacted by man, and it is likely that otherwise comparable estuaries would be normally populated with quite different species assemblages in different parts of the world, but major community characteristics and relationships should be essentially the same except as influenced by environmental stress. If the latter are predominated by anthropogenic factors, these should be apparent with the proper choice of environments.

The geologic record may also provide information useful in detecting and interpreting long-term changes in estuaries, though the technique has been used sparingly to date. Varved sediment cores in certain parts of Chesapeake Bay, dated by pollen and isotope techniques, have revealed changes in phytoplankton community structure, from centric to pennate diatoms, that may be related to historical cultural changes in the region (e.g., colonization, industrialization, intensive farming with the use of inorganic fertilizers). More work of this kind in estuaries should be encouraged, though it is recognized that geologic chronology in soft estuarine sediments, often heavily worked by the benthic biota, is not easy to determine. Equally difficult to interpret, however, is the significance of the ecological changes revealed by these techniques. The change from centric to pennate diatoms, referred to in the example cited above, may be an indication of progressive enrichment of the estuary, but clear evidence for such a relationship is lacking.

Fish scales in varved sediments off the California coast revealed that the disappearance of the sardine in recent times and its replacement by the anchovy, an event sometimes blamed on overfishing or other human activity, has occurred many times in the past, well before anthropogenic influences. These and perhaps other preserved indicators of ecosystem change should be looked for in suitable estuarine deposits.

ESTUARINE HEALTH AND HEALTH INDICATORS

Contrary to one of the more commonly held attitudes of
the day, change in itself may be neither good nor bad.
Changes in productivity, species diversity, or other
parameters of ecosystem structure or function can be used
in reference to the nominative state (which incorporates
natural variations) as described by Odum et al. (1979).
One can use the state variables of a natural system as a
means of comparison. It should be possible to define the
state variables and the nominative state in assessing
estuarine health and its natural variations. However,
change is often referred to some human value concept, and
what change is beneficial to one person may be bad for
another. For example, the best conditions for certain
kinds of commercial fishing may require plankton blooms
dense enough to discolor the water and render it unpleas-
ant for various aesthetic or recreational purposes.

The problem that the environmental manager faces is
to detect early indications of deteriorating health of an
estuary before serious damage occurs. However, early
warning indicators of conditions leading to such catastrophic events are not always so obvious.

The monitoring of water, sediments, and organisms for
toxic trace contaminants is the most common approach in
use today. If the concentrations of heavy metals or toxic
organics approach those that have been found (by bioassay)
to be harmful to the biota or to man if he eats the organisms, it may be possible to locate and eliminate the
source of the contaminant before serious damage occurs.

Chemical and physical processes of both the substance
and the environmental medium, the estuarine system, are
important in identifying pathways of potential transport.
Such relationships as distribution of a material between
soil and water, between air and water, within biological
components, and the potential of transformation products
should be determined To understand how a substance is
distributed within an ecosystem, we must determine its
affinity to the various components of the system. With
such understanding, transformation and degradation rates
for a substance can be integrated with fugacity and partitioning equations (MacKay, 1979; McCall et al., 1981)
to provide information regarding the expected concentrations in various components of the estuarine system.

Because of their notorious ability to concentrate a
host of toxic material--metals, organics, pathogens--
mussels have been used as monitors and whole-water bio-

assay organisms for coastal waters around the world. The "mussel watch" program is not, however, widely in use in estuaries and is, indeed, not a satisfactory approach for many estuaries or parts of estuaries whose waters are not suitable for the bivalve to live.

It is not sufficient, however, to look only for identified toxic materials in the environment, for those that are recognized hazards and that can be and actually are being monitored are but a few of the anthropogenic additions to estuaries. Further, toxic levels are known and standards set for only a few of those. A more useful and safer approach would be the routine monitoring of the health of the organisms themselves. Physiological parameters such as growth, feeding, oxygen consumption, fecundity/reproduction, osmotic/ionic regulation, hematology, nitrogen balance/excretion, and heartbeat/ventilation are all important indicators of organismal health, but baseline levels and normal steady-state conditions are not yet well known for most estuarine species. Carrying out the experimental measurements *in situ* in the environment to be monitored poses difficult logistic problems.

Another approach, perhaps more practical, is to monitor one or more biochemical variables that also provide evidence of environmental stress. Some of those that are being studied for this purpose are adenylate energy charge ratio, lysosomal changes, heavy-metal binding protein production, mixed-function oxygenase induction, and steroid metabolism change. Techniques for measuring these indices and the quantitative relationship of their change to various environmental stresses are still in the early stages of investigation, and adequate baseline data have not been established. Further work in this area should be encouraged.

Disease is a well-recognized, often critical problem in the artificial cultivation of marine finfish and invertebrates. It has, however, received much less attention and recognition in the natural environment with the exception of a few commercial species, such as oysters, utilization of which are usually also considered as a form of mariculture. Are such mariculture sites centers for the spread of infectious disease to native populations? What is the role of introduced exotic species in this regard? Are any of the infectious diseases of fish and shellfish in fact transferrable to man, as has been suggested? What factors affect the ecology of vibrios (e.g., *Lymphocystis*), and what is the role of pollution in their distribution in estuaries? Why are Pacific Coast estuaries epicenters of skin papillomas in flatfish? Is this an infectious disease? Why does it occur only in young? Is there a carrier or secondary host for the disease?

Fin rot/fin erosion disease is now known to occur in European, Southeast Asian, and Japanese estuaries as well as the larger urbanized estuaries of North America. Although it appears to be correlated with wastewater discharges, its basic cause and etiology are still unknown. Does this disease ever occur naturally? What is its relationship to water and sediment chemistry?

These are but a few of the unanswered questions in the challenging new field of marine pathology. Particularly important in this area is the relationship between disease and environmental stress, both natural and that caused by pollution, and there is no environment more appropriate for such study than that of the urbanized estuary.

There are other, much more subtle influences in estuaries that may significantly affect the health and survival of organisms. It has been recognized for at least three decades that coastal and estuarine water at certain times of the year are unsuitable for culturing phytoplankton and a variety of invertebrate larvae. This problem has been experienced by culturists working in such scattered areas as Woods Hole, Narragansett Bay, California coastal waters, and the English Channel. The phenomenon has never been satisfactorily explained nor the causative agent identified. Virtually unobservable in environmental studies, it may nevertheless be a major factor affecting the year-class success of a given species whose larvae may be in a critical developmental stage during such episodes. This is an area in which practitioners of mariculture and estuarine ecology could profitably join forces for the solution of a perplexing problem of interest and concern to both.

Not all environmental stress in estuaries is caused by toxic or inhibitory conditions or additions. The nutrients in domestic and certain industrial wastewaters and in agricultural runoff actually stimulate the growth of plants, as does the enrichment of waters by natural processes such as upwelling. Beneficial in moderate amounts and under some circumstances, since phytoplankton are the primary producers of organic matter responsible ultimately for virtually all other aquatic life, excessive enrichment or "eutrophication" may lead to conditions that are ecologically catastrophic as well as aesthetically displeasing.

Criteria for early warning of the deteriorating health of estuaries from excessive nutrient enrichment are usually considered to be the availability and concentrations of the nutrients themselves and their increase in the water, but such data may be misleading. Nutrients that are growth limiting are usually assimilated as quickly as they become available and reside in the tissues of the biota. Those that are not growth limiting may remain in the dissolved state in the water but are not, then, directly relatable to "eutrophication." A more reliable index of "eutrophication" may be the direct increase of phytoplankton or other plant life (e.g., seaweeds or higher plants, in some cases). However, the relationship between nutrient input, uptake kinetics by the phytoplankton and other autotrophs, and primary production and standing stocks of the plant communities and the herbivores that feed on them are an extremely complex series of interactions for which there are no simple indices.

One of the complicating factors is the identification of the rate-limiting nutrient(s) in estuaries. This is generally agreed most often to be phosphorus in the freshwater environment and nitrogen in the ocean, but it may be either in estuaries and may vary temporally and spatially in a given estuary. If runoff from fertilized farmland is the principal source, the phosphorus often becomes bound to the soil and becomes limiting in the estuary. If wastewater is the source, it is usually deficient in nitrogen relative to phosphorus and nitrogen becomes the growth limiting factor. Thus sources, routes, and rates of nutrient input are important basic data for the study of each individual estuary.

There are growing indications, however, that gross nutrient input is not the key to the onset and development of the unhealthy characteristics of estuaries usually associated with the term "eutrophication." As noted above, stimulation of the right kind of phytoplankton (e.g., centric diatoms, certain flagellates) leads on to increased production of pelagic and benthic finfish and invertebrates, including those of direct value to man. Upwelling areas that support the most productive fisheries on Earth are no less "eutrophic" by any definition of the term, and estuaries that are enriched by incursions of subsurface offshore water via a two-layered circulation are usually highly productive, healthy ecosystems. Enrichment from other sources such as wastewater or other anthropogenic activities, however, often results in blooms of nonmotile green and blue-green algae, and sometimes toxic flagellates, that are not readily utilized as food by higher trophic levels and that accumulate in the water or on the bottom where they create undesirable aesthetic and damaging ecological effects. The same major nutrients may be added in both cases, but their chemical speciation (ammonia, urea, or dissolved organic nitrogen in sewage versus nitrate in deep ocean water), the presence or absence of other nutrients such as silicate that is required by diatoms, the presence of accompanying toxic contaminants such as copper in sewage that inhibits diatoms and favors that of other algae, and the concurrence of enrichment with low salinity and high temperature at the heads of estuaries, where discharges are often located, all may favor the growth of undesirable versus desirable species of algae.

Thus "eutrophication" is a qualitative as well as a quantitative characteristic of estuaries, resulting from a number of interacting factors of which the input of major nutrients is but one. The study of these factors and their interactions is therefore an important area of research needed for understanding and diagnosing the early stages of one of the major causes of the deterioration of estuarine health.

REFERENCES

Cairns, J., Jr. (1982). Predictive and reactive systems for aquatic ecosystem quality control, in *Scientific Basis of Water-Resource Management*, Geophysics Study Committee, National Research Council, National Academy Press, Washington, D.C., pp. 72-84.

Dow, R. L. (1964). A comparison among selected marine species of an association between sea water temperature and relative abundance, *J. Cons., Cons. Perma. Int. Explor. Mer 28*, 425-431.

Heinle, D. R., C. F. D'Elia, J. L. Taft, J. S. Wilson, M. Cole-Jones, A. B. Caplins, and L. E. Cronin (1980). *Historical Review of Water Quality and Climatic Data from Chesapeake Bay with Emphasis on Effects of Enrichment*, report submitted to U.S. Environmental Protection Agency, Chesapeake Bay Program.

MacKay, D. (1979). Finding fugacity feasible, *Environ. Sci. Technol. 13*, 1218-1223.

McCall, P. J., D. A. Laskowski, R. L. Swann, and H. J. Dishburger (1981). Environmental partitioning, in

Testing for Effects of Chemicals in Ecosystems (Appendix A), Environmental Studies Board, National Research Council, National Academy Press, Washington, D.C., pp. 87-92.

Nichols, F. H. (1973). A review of benthic faunal surveys in San Francisco Bay, *U.S. Geological Survey Circular 677*, 20 pp.

NRC Environmental Studies Board (1981). *Testing for the Effects of Chemicals on Ecosystems*, National Academy Press, Washington, D.C., 98 pp.

Odum, E. P., J. P. Finn, and E. H. Franz (1979). Perturbation theory and the subsidy-stress gradient, *Bioscience 29*, 349-352.

3   VARIABILITY OF CIRCULATION AND MIXING IN ESTUARIES*

INTRODUCTION

This chapter provides a brief review of *some* of the important recent developments in our understanding and changes in our perceptions of physical circulation and mixing processes within estuaries. It also presents suggestions concerning future research needs in estuarine dynamics. Because much of the current interest in estuaries has focused on their biological, chemical, and geologic features, more interdisciplinary study of estuaries is called for, and the following discussions tend to emphasize those advances in kinematics and dynamics that are more directly relevant to the description of distributions of biomass, chemical constituents, particulate matter, and pollutants. One objective here is to show through examples that developments in our understanding of the physical forcing and response characteristics of estuaries provide us with at least some of the tools needed to deal with complexities associated with the biological, chemical, and suspended matter distributions in estuaries.

*This chapter was largely developed by a subpanel on circulation and mixing, which consisted of the following: Robert E. Wilson, State University of New York, Stony Brook, *Subpanel Leader*; William Boicourt, The Johns Hopkins University; Malcomb Bowman, State University of New York, Stony Brook; Leonard Haas, Virginia Institute of Marine Science; Takashi Ichiye, Texas A&M University; Donald W. Pritchard, State University of New York, Stony Brook; Maurice Rattray, University of Washington; Jay L. Taft, The Johns Hopkins University; and Mary G. Tyler, University of Delaware.

Another objective is to suggest that interdisciplinary research can serve not only as a platform for evaluation and refinement of existing ideas concerning estuarine circulation and mixing processes but also as a stimulus for significant advances into dynamics and kinematics.

RECENT DEVELOPMENTS AND RELATED RESEARCH NEEDS IN ESTUARINE DYNAMICS

Estuaries are typified by freshwater river and groundwater inputs and saline source waters that require a net outflow through an estuary and provide a longitudinal density gradient to drive a vertical circulation given sufficient estuary depth. Many, if not most, estuaries have appreciable tides and tidal currents that in addition to creating oscillatory variations in properties represent the dominant energy source for turbulence and mixing and contribute to the nontidal circulation through nonlinear effects. Important modifying effects can be provided by surface wind stress over the estuary and by sea-level variations at the mouth produced by meteorological forcing of the adjacent coastal waters.

All the above forcing mechanisms may exhibit temporal and/or spatial variability. Time scales range from geologic periods to seconds, but the range of most interest in terms of estuary-wide response usually lies between the annual and the tidal period. The event time scale of several days associated with meteorological disturbances is likely to result in the most complicated estuarine responses, for it is both broad banded and in the range of the natural time scales for many estuaries.

The spatial scales of variation of forcing mechanisms other than wind are fixed by the boundaries of the estuary, although even here the density variations at the mouth of deep estuaries may exhibit important spatial structure. The important spatial scales may range from the order of a kilometer to hundreds of kilometers.

The estuarine circulation processes of largest spatial scale and longest time scale is the "mean" or nontidal circulation, which is composed of components derived from the net river runoff, the longitudinal density gradient, laterally varying depth-dependent friction, the surface wind stress, and the characteristics of the tide itself; other processes governing the distribution of variables can be categorized as either stirring or mixing, the former resulting from resolved or deterministic motions

and the latter from unresolved or statistically described
motions. The former would include nontidal longitudinal,
lateral, and vertical circulations, and the latter would
include bottom-generated turbulence produced by tidal
currents, mean-flow shear-generated turbulence, boundary
mixing from density currents or internal wave breaking,
and shear and breaking of surface waves generated by the
wind.

The foregoing circulation and mixing processes respond
to the temporal and spatial variability of the forcing,
and, in addition, the mixing and momentum flux processes
produce motions of shorter time and spatial scales. While
these smaller scales may not be of interest in themselves
as part of the overall estuarine description, they must
be investigated before their effects on the larger-scale
processes can be predicted in terms of some suitable
parameterization.

In a selective review of research in the United States
on the dynamics of motions in estuaries, Carter et al.
(1979) point to three of the most important research
advances as an appreciation of

1. The variability in time and space of the nontidal
motion in estuaries.
2. Nonlocal forcing of nontidal variations in water
level and motion in estuaries and other coastal water
bodies.
3. The effects of the nonlinear field acceleration
terms on the tidal averaged dynamics of estuarine
circulation.

This abbreviated list, which emphasizes the importance
of temporal and spatial variability in both physical
forcing and response characteristics, can certainly be
expanded. In addition to temporal variability in nontidal
circulation associated with both local and nonlocal meteorological forcing, Elliott et al. (1978) demonstrated the
importance of forcing at the head of an estuary due to
time-variable river discharge in producing temporal variations in nontidal circulation and in the distribution of
water properties.

As far as spatial variability in circulation patterns
is concerned, there is increased awareness of the importance of dimensional and topographic controls. There is
substantial evidence that our usual two-dimensional
(vertical and longitudinal) simplification of estuarine
circulation is inadequate and that a nonnegligible portion

of the longitudinal transport is associated with effects
in the cross-estuary (lateral) direction. While the lower
layer in a two-dimensional regime is topographically
limited by channels, the upper layer can be complex, exhibiting jet flows, stationary eddies, and sizable lateral
shears. The rotation of the Earth can cause significant
lateral gradients in the larger estuaries. Topographically driven secondary flows have been well documented in
rivers, but little is known about these circulations in
stratified tidal estuaries.

In addition to this relatively persistent spatial
structure, which is associated with irregular topography,
there is increasing evidence for the existence of small-
scale and often short-lived fronts and related spatial
features and for their possible biological consequences
(Figure 3.1).

Temporal and spatial variability have so far been
discussed primarily in connection with nontidal circula-
tion--an advective process. The importance of both temporal and spatial variability can, however, also be
considered in relation to mixing processes. Here we will
consider specifically tidal mixing and its effect on
water-column stability. In this case temporal variability
is associated primarily with spring-neap variations in
mixing intensity and spatial variability in mixing intensity within the estuary.

Tidal mixing intensity can be represented rather simply in terms of bulk parameters, tidal current amplitude,
and mean depth. The more general problem of representing
turbulent transport of mass and momentum remains a fundamental problem in estuarine dynamics and kinematics of
considerable practical importance. Fluid dynamicists
refer to this as the closure problem in turbulence; mete-
orologists or oceanographers may call this "parameterization"; and numerical modelers call it "subgrid
simulation."

The vertical turbulent flux of momentum and mass
results from correlation of the vertical fluctuation
velocity with the horizontal fluctuation velocity or
concentration fluctuation. In contrast to the horizontal
fluctuation velocity and the concentration fluctuation,
generation of the vertical fluctuation velocity, itself,
seems to be caused by different mechanisms near the surface, in the mid-depth and near the bottom of estuaries.
Near the surface, the vertical fluctuation velocity may
be mainly generated by the wind-generated surface waves
and by the shear in the near-surface flow. In the mid-

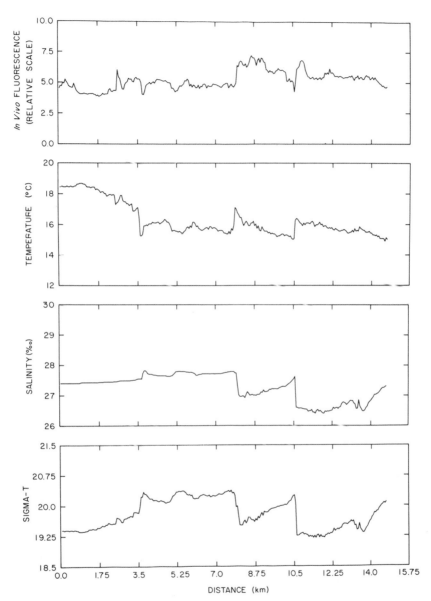

FIGURE 3.1 Small-scale structure in water properties
at 1 m along a horizontal transect in Long Island Sound
from 40° 59' N, 73° 04" W to 41° 07" N, 73° 05' W on 22
June 1977. Fronts at approximately 7.5 km and 10.5 km
are from remnant river plumes; at 3.5 km is a stationary
tidal front off Long Island.

44

FIGURE 3.2 Bottom topography in a 60-km-long reach of Great South Bay, a bar-built estuary the south shore of Long Island, New York. MLW depth is 1.3 m; dotted isobath is 1 m.

FIGURE 3.3 Numerical simulation of Lagrangian drift for a total of ten water parcels for a period of 30 days in the reach of Great South Bay shown in Figure 3.2.

depth the energy source of the vertical fluctuation velocity may be the current shear and the internal waves. Near
the bottom, the topographic features combined with the
current shear may be the main cause of the vertical
fluctuation velocity.

Horizontal turbulent transports of mass and momentum
tend to be associated with variations in velocity and
concentration with time and space scales larger by orders
of magnitude than their vertical counterparts.

We have so far emphasized appreciation of temporal and
spatial variability in circulation and mixing processes in
connection with important recent developments. Of possibly even more fundamental importance is the increased appreciation for the utility of a Lagrangian (with a moving
coordinate system) description of the motion within tidal
areas such as bays and estuaries rather than a Eulerian
(with a fixed coordinate system) description, especially
when considering the net displacement over a tidal cycle
of dissolved and suspended materials. The movement of
dissolved and suspended materials within an estuary is
inherently a Lagrangian process, and for many purposes the
Lagrangian trajectory is the representation most useful
for understanding biological, chemical, and sedimentary
processes and more relevant to the distributions of biomass and chemical constituents.

A problem of special importance is the residual nontidal Lagrangian drift of a particle within an estuary in
which tidal currents are spatially nonuniform and which
is forced both locally and nonlocally at subtidal frequencies. In this case the residual Lagrangian drift will be
strongly affected by an interaction between the spatially
nonuniform tidal stream and the spatially and temporally
variable (at subtidal frequencies) residual Eulerian
velocity field.

Numerical modeling can provide a focus for many of the
physical processes described above. To illustrate this
point, we consider some numerical simulations of the
depth-mean horizontal circulation patterns within Great
South Bay (Figure 3.2), a bar-built estuary on the south
shore of Long Island. The bathymetry is shoal (average
depth 1.3 m, relative to mean low water) and spatially
complex. Its surface area-to-volume ratio makes it highly
susceptible to local wind forcing, and because it communi-
cates directly with the adjacent continental shelf through
four inlets, it is forced by both tidal and subtidal sea-
level fluctuations on the shelf. Figure 3.3 shows a
limited number of Lagrangian trajectories for a 30-day

numerical simulation with a finite-element model. These
simulations were performed with the objective of tracing
the time history of clam larvae within the bay during that
time they are suspended within the water column. They are
presented to emphasize the following basic points:

1. Realistic simulations must be of sufficient length
to cover the primary flow variability due to local meteorological forcing, fluctuations in freshwater inflow, and
forcing by subtidal sea-level changes on the adjacent
shelf.
2. Simulations must be of sufficient resolution to
represent adequately effects due to variations in bathymetry and to be of value to users concerned with transport
processes. In the simulations shown, for example, the
Lagrangian trajectories are spatially very complex and
simulations on the coarse grid would have provided a poor
representation of the structure of the "true" residual
drift field.

The residual nontidal Lagrangian drift as represented
by these trajectories is dramatically different from that
which might be inferred from the residual Eulerian velocity field and even from the sum of the residual Eulerian
field and the Stokes velocity field.

PROCESS-ORIENTED INTERDISCIPLINARY RESEARCH IN ESTUARIES

The nature of the developments in our basic understanding
of physical circulation and mixing processes, more specifically, in our appreciation of the space and time scales
involved, has enhanced our ability to identify and
describe specific physical processes responsible for
maintaining biological and chemical distributions and for
producing temporal and spatial variability in these
distributions. The following discussions are presented
to show, by example, that interdisciplinary research
involving the study of physical processes controlling the
distributions of biomass and chemical constituents can
provide insight into the nature of the physical processes
that would otherwise be unobtainable, or nearly so.
A first example relates to effects associated with
spatial variations in the intensity of tidal mixing and
stationary tidal fronts in estuaries. Within shallow
estuaries spatial variations in mixing intensity can give
rise to changes in water-column structure from mixed to

well stratified in a few hundreds of meters or less.
These variations are important to marine productivity
since the same physical driving mechanisms that control
the stratification also determine the availability of
light and nutrients to the phytoplankton.

One approach to the problem of predicting vertical
stratification in tidally dominated shallow seas is
through the concepts of mixing energetics. During the
heating season, there is a bouyancy flux through the sea
surface due to the insolent radiation. This leads to
warming of the subsurface layers and a tendency of the
water column to stratify, lowering the potential energy
of the column relative to the mixed state. Opposing this
is the generation of bottom turbulence from tidal stream-
ing. Simpson and Hunter (1974) showed that this dynamic
imbalance can be expressed through a stratification index,
h/u^3 (h is water depth and u is mean spring tidal current amplitude). The value of this index determines
whether a coastal ocean region will remain mixed throughout the summer or whether at some stage of the heating
season, onset of stratification commences and a thermo-
cline develops.

A study was initiated in 1978 in Long Island Sound to
see if a similar approach might prove fruitful in a partially mixed estuary when the observed stratification was
almost entirely due to the two-layered gravitational
circulation patterns, i.e., where the buoyancy flux into
a given area was due to *horizontal* advection rather than
a vertical input of buoyancy due to insolation. The
Simpson-Hunter (1974) theory was extended to an estuarine
environment (Bowman and Esaias, 1981) and was shown that
as long as the ratio of upstream to downstream mean salinity was close to unity (equivalent to stating that the
horizontal buoyancy flux is constant along the axis) then
the h/u^3 index should be applicable.

Results of an extensive mapping cruise made in September 1978 in Long Island Sound are shown in Figure 3.4,
where bulk stratification (σT difference bottom to top
divided by depth) is used as a measure of water-column
stability.

Contours of h/u^3, derived from current measurements
made by the National Ocean Survey, are illustrated in
Figure 3.5. There is a remarkable similarity in the two
maps. The $\log_{10} h/u^3 = s = 1.25$ to 1.5 contours appear to define accurately the limits of mixed waters in
the eastern and central regions. The main eastern tidal
channel is flanked by high s (>2) regions of stratified

48

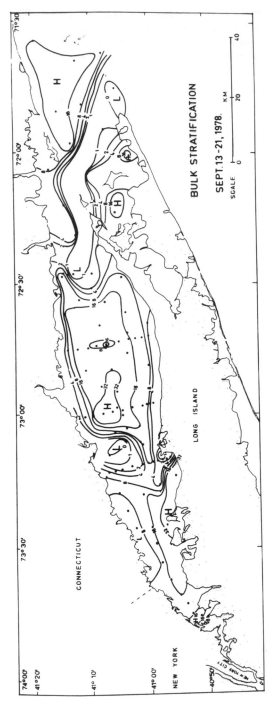

FIGURE 3.4 Bulk stratification (σ_T/h) map of Long Island and Block Island Sounds, September 13-21, 1978 (from Bowman and Esaias, 1981).

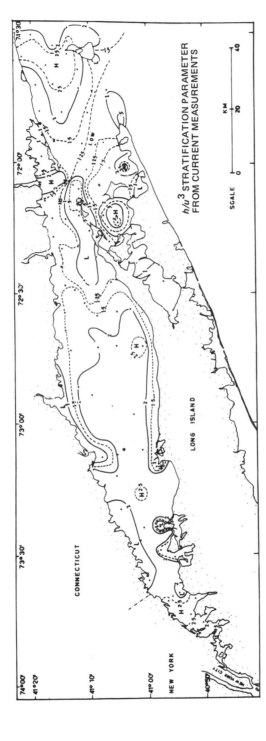

FIGURE 3.5 The h/u^3 stratification index derived from NOS current measurements and bathymetric charts (from Bowman and Esaias, 1981).

water. The central, stratified basins of Long Island
Sound appear as a region of high s, with coastal mixing
zones paralleling the shoreline and a major transverse
front located at its eastern limits.

The distributions of phytoplankton observed on several
sampling cruises are clearly related to these patterns of
tidal mixing and stratification. Further research is
needed to understand the temporal response of phytoplankton to these spatial patterns of mixing and stabilization
and to test the usefulness of the method in other partially mixed estuaries.

A second example relates to effects associated with
temporal variations (spring-neap) in the intensity of
tidal mixing within partially mixed estuaries. Recently
it has been recognized that some estuaries oscillate
between stratified and destratified conditions on a fortnightly cycle that correlates with neap and spring tides,
respectively. For example, the cycle has been observed
in a variety of estuaries including the James, York, and
Rappahannock Rivers (Haas, 1977), the St. Lawrence (Sinclair, 1978), Aulne and Gironde estuaries (Allen et al.,
1980), Charleston Harbor (Wastler and Walter, 1968), and
possibly Puget Sound (Winter et al., 1975) and Saanich
Inlet (Takahashi et al., 1977). The stratification-
destratification process encompasses the entire estuary,
not just a small segment, and affects the fundamental
hydrographic-circulation condition of the estuary. The
time scale of the cycle is long enough for significant
chemical and biological responses to occur. In the York
River estuary the stratification-destratification cycle
is strongly expressed and is predictable (Haas et al.,
1980; Figure 3.6).

Other than its association with the neap-spring tidal
cycle, little is known about the hydrographic processes
causing the observed oscillation between stratified and
homogeneous conditions. However, the fact that the
stratification state of an estuary can be altered so
dramatically on such a short time scale suggests that
further research on this process will significantly advance our basic understanding of the hydrodynamics of
estuarine systems in general.

Other aspects of the estuarine ecosystem are also affected by the stratification-destratification process. In
the York River dissolved oxygen and inorganic nutrients,
both important water-quality parameters, have been shown
to vary systematically with the stratification state
(D'Elia et al., 1980; Webb and D'Elia, 1980). During

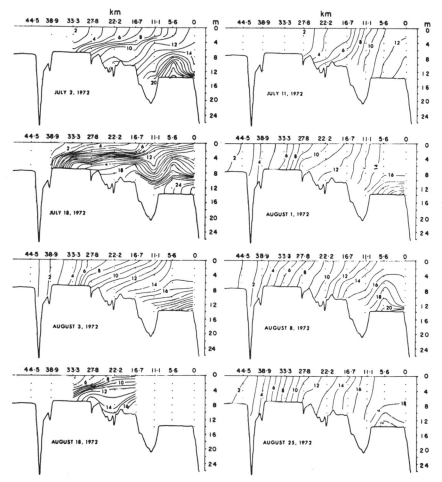

FIGURE 3.6 Vertical sections of salinity along the longitudinal axis of the York River Estuary showing the variation
in water-column stratification associated with spring-neap
variations in tidal current (from Haas *et al.*, 1980).

stratified conditions in the summer months, water below
the halocline becomes simultaneously anoxic and enriched
in both ammonium and phosphate, both conditions apparently
the result of benthic remineralization. This condition
is altered fortnightly by the mixing of oxygen-rich, nu-
trient-poor surface waters with oxygen-poor, nutrient-rich
deep waters. Thus, spring-tide associated destratification events serve to replenish periodically oxygen to the

deep water and nutrients to the surface waters. It is apparent, therefore, that the spatial and temporal variation in both oxygen and nutrient concentration is directly related to the stratification-destratification cycle and that this process may have a significant impact on biological processes of the system.

The last several years have witnessed a growing awareness of the episodic nature of phytoplankton processes and their relationship with physical processes in the oceans (McGowan and Hayward, 1978). Analogous events may be occurring in estuaries as well. For example, several investigations noted sequential phytoplankton blooms in estuaries associated with alternating periods of stratification-destratification (Sinclair, 1978; Takahashi et al., 1977; Winter et al., 1975). These observations suggest a direct association between phytoplankton and hydrographic processes in estuaries that heretofore has not been widely recognized. Other biological processes that may be keyed either directly or indirectly to stratification-destratification cycles include the timing of finfish-shellfish spawns, the survival of larval forms, phytoplankton species succession, fish migrations into or out of an estuary, the conservation or "trapping" of larval forms in an estuary, and zooplankton life cycles.

The spatial and temporal distribution of abiotic constituents as well may be regulated by the hydrographic process in question. Allen et al. (1977, 1980) proposed that the longitudinal distribution of sediments (i.e., the turbidity maximum) in the Gironde and Aulne estuaries is regulated by the observed cycle of neap- and spring-tide-associated stratification and mixing in these estuaries. This is an alternative hypothesis to the prevailing concept that estuarine turbidity maxima occur in the zone of net, zero nontidal flow in an estuary.

A third example, presented from the more general perspective of nutrient dynamics in estuaries, points to the importance of a number of longitudinal and vertical mixing processes. These include the longitudinal dispersion of nutrients contained in runoff, the maximization of vertical concentration gradients between water and sediments by horizontal advection, and vertical mixing across a pycnocline.

Nutrients entering estuaries from rivers may behave conservatively during periods of high flow when dilution with seawater is more rapid than phytoplankton uptake. This often occurs when nutrient concentrations are high. Nitrate input to the Chesapeake Bay during spring behaves

conservatively through most of the region north of the
Potomac River. Linear regressions of nitrate versus
salinity for several years in Chesapeake Bay show that
slopes and intercepts change from month to month ex-
pressing source concentration changes. While this
approach is approximate, with judicious choice of coefficients and intercepts estimated from historical data, it
is possible to model some major nutrient inputs nearly as
well as salinity.

Phytoplankton nitrogen demand appears to be satisfied
by ammonium nitrogen (McCarthy *et al.*, 1977; Taft *et al.*,
1978) so that turbulent mixing with low-nitrate ocean
water is the principal cause of decreasing nitrogen
concentration. Analogous situations exist in the Hudson
River estuary, Columbia River, Mississippi River, and
others.

As sediment accumulates, so do nutrients that are remineralized from organic particles in the sediment. The
nutrients dissolved in interstitial waters can diffuse
back into the overlying waters, as long as the concentration of a given nutrient in the overlying waters is less
than that in the interstitial waters. In this way nutrients entering the estuary from the land are augmented by
remineralized nutrients from sediments so that the
instantaneous mass of nutrients per unit volume steadily
increases.

The input rate from sediments varies for different
nutrient species. Ammonium probably diffuses at a constant rate according to concentration gradient. Nitrate
may diffuse upward when the near-surface sediment is aerobic (Taft *et al.*, 1978), but benthic chamber measurements
have shown nitrate loss to the sediments (Boynton *et al.*,
1980). Phosphate diffuses out at oxygen concentrations
of less than about 1 ml/liter. At higher O_2 levels,
phosphate may be bound to iron hydroxides (formation of
ferric phosphate) and remain at the sediment surface.

Temperature-dependent remineralization of nutrients
in the sediments can be approximated from the Arrhenius
rate law expression

$$\ln \frac{r_2}{r_1} = \frac{E_a}{RT},$$

where r_1 is the rate at a given temperature, E_a is the
apparent activation energy, R is the gas constant, and T
is temperature. Sulfate reduction and ammonium reminal-
ization have been treated this way (Goldhaber *et al.*,

1977; Jorgensen, 1977; Aller and Yingst, 1979). Once
released from organic material, nutrients in interstitial
waters may diffuse upward. Under conditions of deep-water
anoxia both ammonium and phosphate diffuse into the overlying water according to

$$J = \phi D_s \left(\frac{\partial C}{\partial z}\right)_{z=0} ,$$

where J is the flux, ϕ is sediment porosity, D_s is dif-
fusivity for nonbioturbated sediment or eddy diffusiv-
ity for biologically disturbed sediments, and $(\partial C/\partial z)_{z=0}$
is the interstitial water-concentration gradient (Berner,
1971). Krom and Berner (1980) determined diffusion coefficients for ammonium and phosphate in sediments as
$D_s(NH_4) = 9.8 \pm 2.0 \times 10^{-6}$ cm^2/sec and $D_s(PO_4)$
$= 3.6 \times 10^{-6}$ cm^2/sec. Thus, estimates for remineralization rates and release rates from sediments can be
made. Advection in the overlying water transports nutrients horizontally away from the sediments, and turbulence
mixes them vertically. In this way gradients are maxi-
mized and diffusive fluxes maintained from interstitial
to overlying water.

At present, estimates for release rates from sediments
are better than estimates of mixing through the water
column. This is because sediment release may be adequately modeled as a steady-state process, whereas circulation
of the overlying water is not in steady state. However,
water circulation is primarily responsible for bringing
phytoplankton and nutrients together.

Under conditions of strong stratification, nutrients
(predominantly ammonium and phosphate) accumulate in the
lower layer. The pycnocline may or may not coincide with
the bottom of the euphotic zone. Phytoplankton with
limited mobility are thus virtually isolated from direct
contact with the deep nutrient pool and depend entirely
on physical mixing across the density discontinuity for
new nutrients to support biomass increases. Dinoflagel-
lates are an exception because they may migrate vertically
to the boundary and take up the nutrients available there.
Phytoplankton with limited mobility can be exposed to nutrients in two ways. First, new nutrients from below the
pycnocline can be mixed or diffused up or the organisms
can be mixed downward. In actuality, both probably occur,
but rates are poorly defined. Thus the biomass or standing crop of phytoplankton is closely linked to the intensity of mixing within the upper layer and to the intensity

and characteristics (e.g., temporal variability) of mixing across the pycnocline. There is a need to further examine mixing rates in estuaries with tracers other than salinity and density. Errors in estimating diffusivities and transient mixing events may have significant impact on biological and chemical parameters but remain undetected by steady-state mixing or diffusion calculations.

A final example is presented to show that study of physical processes controlling biological distributions can serve to demonstrate the integrated effects of the interaction of transport processes operating over a broad range of space and time scales. It suggests also that biological species can be effectively used as a tracer to distinguish specific advective and nonadvective transport processes responsible for producing observed distributions.

Tyler and Seliger (1978, 1980) described the spatial distribution of plankton, primarily the dinoflagellate *Prorocentrum*, within the Chesapeake Bay in relation to the regulating physical processes. They described the seasonal distribution and transport of organisms by the "mean" nontidal density driven flow. Seaward flowing populations within the surface layers are transported into the bottom waters at frontal regions near the mouth of the Bay where the pycnocline intersects the surface. Algae are accumulated within the frontal region, giving rise to a surface patch as well as to concentrations entrained along the frontal interface (Figure 3.7A). A freshet from the northern Bay then causes this vertical discontinuity to become a horizontal pycnocline along with the associated algae (Figure 3.7B). These populations are then transported up estuary at depth by the nontidal flow and returned to the surface layers in the northern Bay. This process is inferred from a series of vertical sections along the longitudinal axis of the Bay (Figure 3.8): Figure 3.8B shows the low stratification and surface concentrations in the lower Bay in mid-winter; in early spring there is an onset of stratification and a transfer of algae to the bottom waters (Figure 3.8C); during the late spring organisms are transported to the northern Bay under high stratification (Figure 3.8D); by summer the delivery of algae to the northern Bay is complete (Figure 3.8E).

In addition to relating concentration distributions to "mean" nontidal circulation patterns, Tyler and Seliger (1978, 1980) considered low-frequency variability in algae distributions and the reponse to variability in physical forcing. They described the temporal variability in the

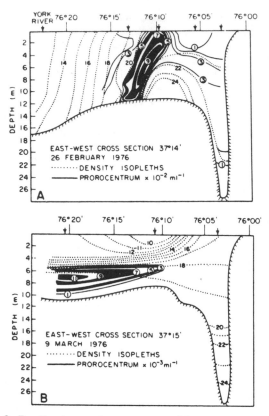

FIGURE 3.7 East-west cross section across the Chesapeake Bay looking northward (at York River entrance) showing displacement of *Prorocentrum* originally accumulated at frontal interface A downward beneath pycnocline B to produce a subsurface maximum (from Tyler and Seliger, 1980).

first appearance of algae accumulations along the frontal region in the lower Bay (February through March) and subsequent appearance of subsurface concentration maxima (February through June). They related variability in the first appearance of algae blooms in the upper Bay to the apparent variability in the intensity of subsurface density-driven transport that was produced, in turn, by variations in freshwater inflow.

Tyler and Seliger also showed that certain aspects of the spatial variability in the distributions of planktonic organisms are strongly related to the spatial structure of circulation and mixing patterns. In addition to algae

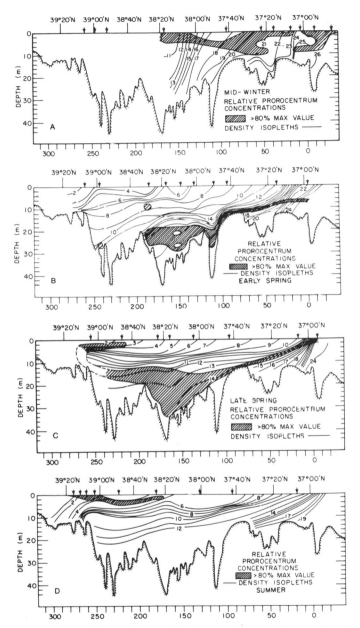

FIGURE 3.8   Series of vertical sections along the longitudinal axis of Chesapeake Bay showing the northward transport of depth of *Prorocentrum* by the large-scale density driven flow (from Tyler and Seliger, 1978).

58

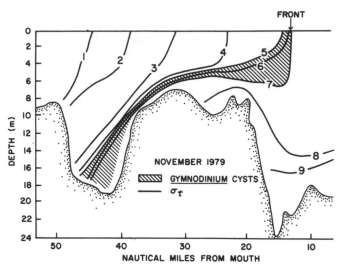

FIGURE 3.9 Longitudinal section of Potomac River showing the relation of longitudinal distribution of benthic Gynmodium cysts and intercept of halocline with bottom.

accumulations within the frontal regions in the lower Bay, they demonstrated, for example, that the patchy distribution of benthic beds of resting cysts of the red tide organism *Gymnodium neboni* in the Potomac Estuary can be delineated in the longitudinal direction by the extent of salt wedge penetration (Figure 3.9); in the lateral direction highest cyst concentrations were found where the pycnocline intersects the bottom on the south side of the estuary (Figure 3.10). They also discussed evidence for the importance of the smaller-scale frontal processes

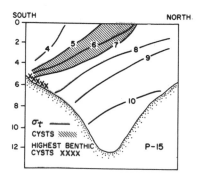

FIGURE 3.10 Lateral section across the Potomac River showing the relation of lateral distribution of benthic Gymnodium cysts and intercept of halocline with bottom.

superposed on the larger-scale circulation patterns to the spatial distribution of plankton.

More general questions that appear answerable by this type of interdisciplinary research revolve around topics of patchiness of phytoplankton, bacteria and zooplankton, seasonal succession, productivity and bloom formation, and community structure.

Such research involving examination of biological and chemical constituent distributions should provide considerable insight into mechanisms of transport by both advective and nonadvective processes and into the effects of temporal and spatial variability in these transport processes on the distributions.

REFERENCES

Allen, G. P., G. Sauzay, P. Castaing, and J. M. Jouanneau (1977). Transport and deposition of suspended sediment in the Gironde estuary, France, in *Estuarine Processes*, M. L. Wiley, ed., Estuarine Research Federation, Academic Press, New York, pp. 63-81.

Allen, G. P., J. C. Salomon, P. Bassoullet, Y. du Penhoat, and C. de Grandpre (1980). Effects of tides on mixing and suspended sediment transport in macrotidal estuaries, *Sediment. Geol. 26*.

Aller, R. C., and J. Yingst (1980). Relationships between microbial distributions and the anaerobic decomposition of organic matter in surface sediments of Long Island Sound, U.S.A., *Mar. Biol. 56*, 29-42.

Berner, R. A. (1971). *Principles of Chemical Sedimentation*, McGraw-Hill, New York, 240 pp.

Bowman, M. J., and W. E. Esaias (1981). Fronts, stratification, and mixing in Long Island and Block Island Sounds, *J. Geophys. Res. 86*, 4260-4264.

Boynton, W. R., W. M. Kemp, and C. G. Osborne (1980). Benthic nutrient fluxes in the sediment trap portion of the Patuxent estuary, in *Estuarine Perspectives*, V. S. Kennedy, ed., Estuarine Research Federation, Academic Press, New York.

Carter, H. H., T. O. Najarian, D. W. Pritchard, and R. E. Wilson (1979). The dynamics of motion in estuaries and other coastal water bodies, *Rev. Geophys. Space Phys. 27*, 1585-1590.

D'Elia, C. F., K. L. Webb, and R. L. Wetzel (1980). Impact of hydrographic events on water quality in an estuary, in *International Symposium on Nutrient*

Enrichment in Estuaries, B. J. Neilson, ed., The Humana Press, Inc., Clifton, New Jersey, p. 597.

Elliott, A. J., D. P. Wang, and D. W. Pritchard (1978). The circulation near the head of Chesapeake Bay, *J. Mar. Res. 36*, 643-655.

Goldhaber, M. B., R. C. Aller, J. K. Cochran, J. K. Rosenfeld, C. S. Martens, and R. A. Berner (1977). Sulfate reduction and bioturbation in Long Island Sound sediments: Report of the FOAM group, *Am. J. Sci. 277*, 193-237.

Haas, L. W. (1977). The effect of the spring-neap tidal cycle on the vertical salinity structure of the James, York and Rappahannock Rivers, Virginia, U.S.A., *Estuarine Coastal Mar. Sci. 5*, 485-496.

Haas, L. W., F. J. Holden, and C. S. Welch (1980). Short term changes in the vertical salinity distribution of the York River estuary associated with the neap-spring tidal cycle, in *Nutrient Enrichment in Estuaries*, B. Neilson and L. E. Cronin, eds., The Humana Press, Inc., Clifton, N.J., p. 585.

Jorgensen, B. B. (1977). The sulfur cycle of a coastal marine sediment (Lim Fjorden, Denmark), *Limnol. Oceanogr. 22*, 814-832.

Krom, M. D., and R. A. Berner (1980). The diffusion coefficients of sulfate, ammonium, and phosphate ions in anoxic marine sediments, *Limnol. Oceanogr. 25*, 327-337.

McCarthy, J. J., W. R. Taylor, and J. L. Taft (1977). Nitrogenous nutrition of the phytoplankton in the Chesapeake Bay. 1. Nutrient availability and phytoplankton preferences, *Limnol. Oceanogr. 22*, 996-1011.

McGowan, J. A., and T. L. Hayward (1978). Mixing and oceanic productivity, *Deep-Sea Res. 25*, 771-794.

Simpson, J. H., and J. R. Hunter (1974). Fronts in the Irish Sea, *Nature 250*, 404-406.

Sinclair, M. (1978). Summer phytoplankton variability in the lower St. Lawrence estuary, *J. Fish. Res. Bd. Can. 35*, 1171-1185.

Taft, J. L., A. J. Elliott, and W. R. Taylor (1978). Box model analysis of Chesapeake Bay ammonium and nitrate fluxes, in *Estuarine Interactions*, M. L. Wiley, ed., Estuarine Research Federation, Academic Press, New York.

Takahashi, M., D. L. Seibert, and W. H. Thomas (1977). Occasional blooms of phytoplankton during summer in Saanich Inlet, B. C., Canada, *Deep-Sea Res. 24*, 775-780.

Tyler, M. A., and H. H. Seliger (1978). Annual subsurface
transport of a red tide dinoflagellate to its bloom
area. Water circulation patterns and organism distributions in the Chesapeake Bay, *Limnol. Oceanogr. 23*,
227-246.

Tyler, M. A., and H. H. Seliger (1980). Selection for a
red tide organism. Physiological responses to the
physical environment, *Limnol. Oceanogr. 26*, 310-324.

Wastler, T. A., and C. M. Walter (1968). Statistical
approach to estuarine behavior, *J. Sanitary Eng. Div.,
ASCE SA6, Proc. Paper 6311*, pp. 1175-1194.

Webb, K. L., and C. F. D'Elia (1980). Nutrient and oxygen
redistribution during a spring neap tidal cycle in a
temperate estuary, *Science 207*, 983-985.

Winter, D. F., K. Banse, and G. C. Anderson (1975). The
dynamics of phytoplankton blooms in Puget Sound, a
fjord in the Northwestern United States, *Mar. Biol.
29*, 139-176.

4

SUSPENDED AND DISSOLVED MATTER IN ESTUARIES

INTRODUCTION

The sources of materials to the estuarine and coastal zone are both natural and anthropogenic. The fates of these dissolved and particulate substances, however, are controlled by the general properties of the system: geometry, currents, and organisms. Thus the study of the "physiology" of the system is fundamental to the identification, understanding, and remedying of any "pathology" of the system resulting from human actions.

Dissolved and particulate substances in estuaries, including both inorganic and organic materials, can be supplied by upland drainage, from the continental shelf, from atmospheric fallout, by shore and bottom erosion within the estuary, and by organisms within the system.

*This chapter was largely developed by subpanels on (1) suspended particulate matter and (2) fate of materials, which consisted of the following: (1) Robert B. Biggs, University of Delaware, *Subpanel Leader*; James M. Coleman, Louisiana State University; L. Eugene Cronin, Chesapeake Research Consortium; John Fischer, Clemson University; David H. Peterson, U.S. Geological Survey; and Jack Pierce, Smithsonian Institution and (2) Karl K. Turekian, Yale University, *Subpanel Leader*; Roy Carpenter, University of Washington; Christopher S. Martens, University of North Carolina; Willard S. Moore, University of South Carolina; Scott W. Nixon, University of Rhode Island; James Simpson, Lamont-Doherty Geological Observatory; and Herbert L. Windom, Skidaway Institute of Oceanography.

Suspended materials, as physical particles, can act as
pollutants to organisms, by clogging gills or interfering
with feeding, though most estuarine organisms are adapted
to relatively high concentrations of suspended materials.
Suspended materials can affect the aesthetics of estuarine
waters, as high turbidity is regarded frequently as in-
dicating pollution, at least for water-contact activities.
The character and quantity of suspended solids has a direct effect on the quality and quantity of light reaching
primary producers. Suspended solids serve as both a reaction substrate and as a carrier for a host of inorganic
and organic pollutants, affecting both their form and
fate.

The questions involved in understanding the fate of
materials in the estuarine and coastal zone can be phrased
as follows:

1. What is the riverine flux of dissolved and particulate substances? How does it vary with time in any one
river, and what differences are there among rivers?

2. What reactions occur at the freshwater-seawater
boundary to sequester or release chemical species to the
ocean? What is the role, for example, of salinity changes
and intertidal marshes in regulating the flow of dissolved
species to the ocean?

3. What is the scavenging role of biological activity
in the water column, in the resuspended sediments, and in
the chemical precipitates of manganese and iron released
from sediments?

4. How are sediments redistributed as the result of
tidal currents and episodic storms? What is the net
effect in the coastal zone of estuarine flow and wind-
driven upwelling or downwelling on the accumulation or
exportation of sediments?

5. What is the role of bioturbation in either
releasing chemical species to the water or increasing the
trapping efficiency of the sediment for the deposited
elements?

Although there are many other questions that can be
phrased to detail the answers more precisely, the above
cover the main ones that have been the incentive for much
innovative work done in estuaries in recent years. The
answers to these questions are not entirely obvious.

TRANSPORTATION AND FATE OF REACTIVE CONSTITUENTS
ASSOCIATED WITH PARTICULATE SUSPENDED MATTER

Natural weathering and erosion processes lead to river
turbidity in the form of suspended-sediment load. The
quantity of suspended sediment varies over several orders
of magnitude, depending on the stage of fluvial development. Several small, natural, pristine rivers in the
northeastern United States have concentrations of suspended solids ranging from 1 to 25 mg/liter as compared
with pristine watersheds in the southwestern United States
wherein suspended solids range from 10 to 50,000 mg/liter
(U.S. Geological Survey, 1975).

Land uses such as agriculture, industry, and high-
density housing may increase the susceptibility of land
to erosion. The suspended-sediment contributions to
streams can be highly variable. Contributions to the
suspended-sediment load are known from industrial effluents, sewage-treatment effluents, parking-lot runoff,
atmospheric fallout of particulate materials, and resuspension of materials already deposited. Soil-conservation
practices and both large- and small-scale impoundments,
may serve to reduce these high levels of suspended
materials.

Concentrations of suspended matter in the open ocean
are quite low, and gradients in estuaries typically show
decreasing suspended matter concentrations in a seaward
direction. Residual bottom currents are frequently
directed up the estuary, contribute to the movement of
sands into the mouth of estuaries and tidal inlets, and
may transport near-bottom suspended materials upstream.
Of the two potential source areas for suspended matter to
an estuary, continental areas are much more important than
the sea. Within the land-source area, a complex inter-
relationship between environmental factors determines and
controls the rate and volume of fine materials supplied to
the estuaries.

Estimation of the net flux of material carried by
fluvial processes to the ocean is difficult for a number
of reasons, including the large degree of variation of
properties of interest on a range of time and space scales
within the estuarine zone of transition between freshwater
and seawater. This is especially true for the fine
particles (silt and clay), which accomplish much of the
transport and accumulation of a number of substances of
interest in marine chemistry. Tidal currents can carry
large quantities of fine particles during maximum flow

velocities but may be slow enough during slackwater periods, in restricted embayments, or near the upstream end of two-layer estuarine circulation to effect deposition of a major portion of the suspended load. Resuspension may or may not occur, depending on many factors, and thus net fine-particle estuarine transport often consists of a long series of episodes of resuspension and deposition. Current velocity and suspended particle-mass or particle-size measurements, even over extended periods of time at a number of sampling stations are usually not sufficient to define the net flux averaged over periods of weeks to years.

The most successful approaches to defining net particle fluxes require measurements of water velocities and particle masses for guidance in choosing the most critical sampling locations, but they have depended primarily on budgets deduced from the distribution of "tracers" already in the system. Some of the most effective tracers for defining net fine-particle fluxes in estuaries are radionuclides, both natural and anthropogenic (Turekian et al., 1980). For example, plutonium derived from global fallout is strongly associated with soil particles and moves with fine particles carried in rivers and estuaries or deposited in the sediments. Radioactive cesium and other gamma-ray emitting radionuclides, derived from either global fallout or from low-level releases from nuclear facilities, such as commercial power plants, have also been used successfully to deduce net fine-particles transport and accumulation patterns of fine particles. Other tracers that have been valuable for these purposes include natural radionuclides such as ^{210}Pb, trace metals, and chlorinated hydrocarbons such as polychlorinated biphenyls.

In general, the fine particles carried downstream into an estuary have quite different relative concentrations of various potential tracer substances than of coastal marine particles, a portion of which may move into an estuary with the net upstream motion of the deeper water. Thus sediments accumulating in an estuary may include fine particles derived from both the freshwater and seawater end members. Resolution of the relative and absolute contributions of those two sources to the net fine-particle fluxes and sediment accumulation rates requires considerable effort to explore the differences in end-member tracer properties but is essential to provide a framework for deducing the behavior of other nonconservative substances in an estuary. This is especially true when attempting to deduce the time history of net fine-particle and reactive substance fluxes over periods of years to decades.

An example of the use of chemical tracers in understanding the role of estuaries in modifying the riverine flux of materials to the ocean has been the study of radium and barium in these systems. Five years ago we had little understanding of the way reactions within estuaries modify the fluxes of these elements to the ocean. Now we can make rather precise statements about the reactions to be considered and the expected interactions with other variables within the systems. Studies in the Hudson River-New York Bight, Narragansett Bay, Long Island Sound, Winyah Bay, Chesapeake Bay, and Mississippi River reveal that there is an initial desorption of Ra and Ba from riverine sediments entering the low-salinity reaches of the estuary as well as a continued flux of radium from sediments throughout the estuary. These reactions significantly augment the fluxes of Ba and ^{226}Ra and possibly ^{228}Ra to the ocean.

The transport, reaction, and fate of manganese in estuaries are more complex. Manganese enters the estuary as (1) dissolved Mn^{2+}, (2) coating of MnO_2 on particles, (3) Mn^{2+} adsorbed to particles, and (4) complexes with other dissolved materials, particularly humic acids. Within the estuary, processes occur that affect the concentration and distribution of Mn (Sholkovitz, 1978; Aller, 1980; Elderfield et al., 1981). The most important of these are thought to be (1) flocculation of organic acids, (2) desorption of Mn^{2+} from particulates, (3) burial and reduction of MnO_2 in sediments, and (4) oxidation of Mn^{2+} in the water column. It has been observed that the total dissolved and particulate Mn concentration in the water column in the estuary often exceeds the total concentrations of Mn in the river and ocean end members. The following Mn cycle has been postulated to explain this surplus Mn.

Particulate Mn enters the estuary and settles into anoxic sediments. Here MnO_2 is reduced to soluble Mn^{2+}, which diffuses into the overlying water and moves up the estuary with the incoming saltwater. As it mixes into surface waters, the Mn^{2+} moves toward the ocean in accordance with estuarine circulation. During transport, Mn^{2+} may be oxidized to Mn^{4+} or adsorbed onto particles that settle back to the bottom, where the cycle begins again. This cycle produces a net retention of Mn in the estuary, leading to concentrations.

Although the concept of Mn cycling and the estuary
trap is fairly well established, numerous areas of active
investigation remain. One major problem is to distinguish
chemical speciation (especially Mn^{2+} from Mn^{4+}) in the
various phases. A second problem involves determining the
role of bacteria in geochemical cycles. Mn oxidation and
reduction seem to be mediated by bacteria, although con-
clusive proof of the relative contributions of biological
and inorganic processes remain to be demonstrated. Fur-
thermore, we know little about how floods or storms affect
the system. How rapidly will the cycle be re-established
after major perturbations? What materials are responsible
for complexing Mn? How strong are these complexes? What
other metals or materials are carried through the cycle
because they bind with MnO_2 and are not sequestered in
reducing sediments?

A further example of the use of chemical tracers is
the combination of analyses of trace organic compounds
(aliphatic and polynuclear aromatic hydrocarbons--PAH)
with analyses of radioisotopes in the same sediment and
particulate samples from Pacific Northwest estuaries
(Prahl and Carpenter, 1979; Prahl et al., 1980). ^{210}Pb
fluxes have been used to validate in situ essentially 100
percent trapping efficiency of the type of sediment traps
used to collect sinking particulates for aliphatic and PAH
analyses. If the traps were quantitatively sampling the
vertical fluxes the yearly vertical fluxes of aliphatic
and aromatic hydrocarbons collected in the traps could be
compared with the total annual depositional fluxes in the
bottom sediments 50 m below. This comparison showed the
following: (1) 2-8 times faster recycling of marine than
terrestrially derived aliphatic hydrocarbons and (2)
fluxes of all the PAH except possibly perylene in the
sediments could be accounted for by sinking preformed into
the sediments, with no measurable diagenetic formation or
degradation after burial in the sediments. From the
^{210}Pb-derived sedimentation rates it has been shown that
the large increases in PAH and aliphatic hydrocarbon
concentrations in Puget Sound region sediments began
around 1900, long before major sewage treatment plant and
oil refineries began discharging their wastes into the
Sound in the 1960s.

A final example is the use of the radioisotopes ^{65}Zn
and ^{210}Pb to trace Columbia River effluent and sediments
in the northeastern Pacific Ocean. The ^{65}Zn was artifi-
cially produced at the Hanford reactors. Its distribution
in the adjacent Pacific Ocean showed how the Columbia

River plume responds to seasonal changes in winds (Cut-shall *et al.*, 1971). In summer predominately northerly winds drive the effluent to the south and offshore, while in winter predominately southerly winds drive the effluent northward and along the Washington coast. The highest ^{65}Zn activities in sediments showed that the sediment discharge was primarily transported to the north along the middle of the Washington continental shelf, with 30-40 percent being transported off the shelf. Profiles of activity of the natural radionuclide ^{210}Pb with depth in sediment cores on the Washington shelf (Carpenter *et al.*, 1981) showed that two thirds of the estimated Columbia River discharge of suspended solids accumulates in the north-northwesterly trending midshelf silt deposit heading toward the Quinault submarine canyon (Figure 4.1). This picture of the shelf sedimentation patterns is in agreement with bottom current meter measurements and seabed drifter dispersal patterns.

Scavenging of Chemical Species from Estuarine and Coastal Waters

The scavenging time scales of "reactive" chemical species by suspended particles in the estuarine and coastal zone has been determined by following the distribution of two short-lived thorium isotopes produced in the ocean, ^{228}Th (1.9 yr) produced from ^{228}Ra (5.7 yr) and ^{234}Th (24 days) produced from ^{238}U (4.5×10^9 yr). The mean residence time relative to removal to the sediment has been shown to be dependent on the particle concentration in the water column. In Long Island Sound the residence time of ^{234}Th is about a day. In the New York Bight it increases almost monatomically with distance from shore from a few days at the inner Bight apex to tens of days as the continental shelf break is approached. Thereafter the open-ocean value of hundreds of days becomes operative.

Any chemical species behaving like thorium can be expected to follow suit and thus be found in the sediment pile soon after injection into the estuarine zone. The standing crops of ^{210}Pb and several pollutant-associated metals in sediments of Long Island Sound confirm this expectation.

The introduction of ^{234}Th, ^7Be, ^{210}Pb, and plutonium into coastal sediments permits the determination of bioturbation rates at different levels in the sediment pile and through this an estimate of the final repositories of materials introduced into the estuarine zone. Because of

70

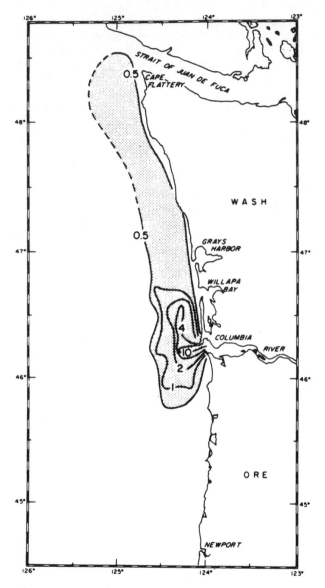

FIGURE 4.1   Isopleths of ^{65}Zn in the uppermost 1 cm of
shelf sediments off the Columbia River mouth (after Cut-
shall *et al.*, 1971). Activity per unit area is expressed
in nanocuries/meters^2.

the role of biological and physical reworking the sediment
accumulation rates in certain estuaries may be difficult
to measure unless long cores (5 to 15 m) are raised,
sufficient to sample below the depths of large animal
burrows (e.g., *Squilla*).

At the same time that an estuary is sampled for trace
elements, particle density, salinity, nutrients, and the
distribution of ^{234}Th should also be made in the water
column and the sediment column as a guide to the efficiency of scavenging and the time scale of homogenization
of surface sediments, especially the clay-size fraction.
This measurement will serve as a valuable guide to the
short-term distribution of other reactive chemical species
whether introduced isotopically (via the air) or point
pollution sources (sewer outfalls or contaminated rivers).

Fate of Particulates--Research Questions

The fate of the particulate matter in estuaries includes
change to dissolved or gaseous forms, burial, or transport
to the sea. Except in short estuaries or in the case of
extreme events, most suspended matter probably takes considerable time to transit the estuarine system or to be-
fall one of its other possible "ultimate" fates. During
the transit time, the material may be deposited, resuspended, ingested, filtered, and partially dissolved, and
its composition and reactivity may vary along the estuarine gradient and with the various estuarine environments.

What is the importance of suspended particulate matter
as a substrate for chemical reactions and reactants? What
are the important interactions of suspended matter with
phosphorus and nitrogen, major and trace metals, and
natural and anthropogenic organic materials? It is fair
to say that, with few exceptions, our understanding of
estuarine particulate matter is rudimentary. We have not
applied advanced, automated techniques to measure particle
size, composition, or concentration. We have chemically
analyzed suspended matter for elemental composition but
have not investigated speciation. We know little about
transformation of chemical species or reaction rates and
can assemble only the simplest predictive models.

How can we measure physical parameters and chemical
processes in the microenvironment (<100 μm)? We should
investigate the advanced "micro" techniques used by micro-
biologists and biochemists for applicability to suspended
matter.

Major research questions concerning the behavior and interactions of suspended materials and their chemical environment include the following:

1. What is the role of the living (bacteria and phytoplankton) portion of the suspended matter in the overall accumulation and concentration of reactive materials? We do not know the relative importance of the biota and the nonliving organic portion of the particulate matter with respect to accumulation and speciation of reactive constituents.
2. Which species of important chemicals react with particulates, and how are they transformed after association? Both the reactivity and toxicity of important chemicals may change after association with particulates. Additionally, inorganic particulate-dissolved interactions may be affected by the mode of attachment of important chemical species to particles.
3. To what extent do clays serve as catalysts for important estuarine chemical reactions, retaining their own character while increasing reaction rates and reaction pathways?
4. What is the relative importance of suspended sediment reactions in the estuarine system as contrasted with those of the dissolved phase or those occurring in association with the benthic environment?
5. What environmental factors (pH, salinity, dissolved oxygen, density gradients) contribute to the partitioning of chemicals between dissolved and particulate phases or between water masses?

ESTIMATES OF THE EFFECTS OF APERIODIC EVENTS ON PARTICULATE MATERIALS

Unusual events can have enormous and important effects on estuarine systems, and one of the areas of greatest impact is on the quantity, composition, distribution, and effects of the particulate material in the system. It is not now possible to predict this impact with sufficient accuracy to protect public interests. That difficulty is evidence that understanding of the effects is seriously weak.

Events that are discussed here under the term "aperiodic" are those that occur rarely and without regularity and that have substantial effects on the individual system involved. Some are natural events, and some are anthropogenic in origin. Variety is great and includes the following:

1. *Floods*--ranging in scale from heavy rainfall in a marsh or small estuary to regional storms with rainfall of frequency of one event in several hundred years.
2. *Human modification of flow*--opening of the Bonnet Carré spillway from the Mississippi River to Lake Pontchartrain or collapse of a major dam are examples.
3. *Faunal or floral shifts*--widespread and rather sudden invasion of oyster beds by serious parasites or predators can remove a large sediment-affecting population. Conversely, extensive invasion of a new area by a filter feeder like *Rangia cuneata* can add a mechanism. Extensive reduction or increase in beds of submerged aquatic vegetation or in marsh stands has large effects on the particulate materials of the estuary.
4. *Spills and releases*--large-scale spills of oil or other chemicals and introduction over time of particle related chemicals like kepone can affect the chemical composition and other attributes of huge quantities of suspended material.

The general nature and magnitude of effects of some of these are known. Floods in the Susquehanna River raise the flow from an average of 40,000 cubic feet per second (cfs) to 650,000 cfs (in 1936) or even 1,000,000 cfs (in 1972). Such flows carry enough sediments into the upper Chesapeake to create new islands and raise the floor of the upper Bay, introduce enough pollutants to threaten useful stocks of animals and plants, carry enough pathogens to threaten public health and some of the estuary's biota, and modify the total hydrographic regime of the Bay (Chesapeake Research Consortium, 1976). Obviously, effects on particulate materials were enormous. Operations of the Bonnet Carré spillway has been observed to kill some species in Lake Pontchartrain, modify the floor of the lake, and, eventually, stimulate vigorous recovery of the biota. Spills of many kinds have been described, although the impacts on particulate materials is rarely well documented. In the James River, several tons of kepone have become associated with particulate matter and deposited throughout the estuary.

The effects of such events on the quantity, composition, distribution, and role of particulate materials is known to be great. This impact cannot be estimated in advance with useful precision, only partially measured after the event. Our ability to predict effects of extreme evemts is weak because we lack knowledge of the following three essential elements:

1. Sufficient baseline or reference information on the normal patterns of circulation, transport, and chemical processes and biotic response to permit quantification of the induced changes.

2. Sufficient description of the changes in chemical and physical content, transport, deposition, and biotic response associated with such aperiodic events.

3. Tested theory explaining observed effects of such events.

There are inherent and serious difficulties in this learning process. The events are usually unexpected, so that research often cannot be planned with assurance and efficiency. Many such events obviously occur during hurricanes, heavy winds, or other conditions involving high hazard for observers and observation equipment. Most interpretations will depend on the availability of good background or baseline conditions, and these are scarce indeed among American estuaries. Finally, knowledge will be dependent on adequate techniques and instrumentation to measure the concentration, composition, and distribution of particulate materials; present methods are far from adequate for these purposes.

CHEMICAL AND PHYSICAL PROCESSES IN THE BENTHIC BOUNDARY LAYER

Modeling the Benthic Boundary Layer

A reasonable end product to the understanding of any estuarine ecological system is the development of a conceptual model, written in quantitative form, in which one has confidence. This has been done for some relatively simple estuarine phenomena. Often the hydrodynamic variables and their associated vertical and longitudinal fluxes can be stated with some certainty. The processes and transformations within the water column related to such items as phytoplankton production respiration and decay, zooplantation grazing, and nutrient uptake and regeneration are understood to a lesser degree. The exchange between particulate and dissolved phases within the water column and on the bottom are poorly understood. *A priori*, it is difficult to develop a convincing ecosystem model in which these latter two processes may be important if we cannot quantify their relationship in the defining model.

This difficulty may arise in the modeling procedures in the following manner. As the benthic fluxes are often not measured directly, they are used as the adjustments to verify the model with observation of distribution in the water column. This adjustment could be taken as a means of indirectly determining the benthic effects, but it should be recognized as such and not as a proof of the model itself. Further, this adjustment factor can, and does, vary considerably from one modeling procedure and comparison with field data to another, which may or may not represent actual variations. In one model we may find a given constituent to be a bottom source, while in another it will be a bottom sink.

Why has this dilemma arisen? Within the water column we can measure fluxes, transformations, and the physical, chemical, and biological variations; we can understand these processes with varying degrees of certainty. On the other hand, bottom and near-bottom processes, of both a chemical and physical nature, are more difficult to observe because gradients are sharp over short vertical distances and because the physical and chemical structure may be modified rapidly and drastically with tidal oscillations and other changes in water movement. Because of this difficulty we simply do not understand them as well as those processes that occur within the water column, and we sometimes omit them from modeling considerations out of ignorance.

We consider it correct to state that processes and transformations relative to the bottom are one of the serious unknowns in our understanding of estuaries.

The zone in the immediate vicinity of the bottom is characterized by high suspended-particulate concentrations and high shear stresses. The processes of sedimentation and resuspension in this zone are highly variable and are influenced by a wide range of physical and biological parameters. An adequate description of these processes requires an understanding of the fluid velocity field, including the turbulent-boundary layer, the chemical and biological characteristics of the particulate (e.g., cohesive properties, organic slimes), and the particle settling characteristics.

The bottom morphology and the texture of the bed material have considerable influence on the processes of settling and resuspension. The range of bottom types found in different estuaries, and even within the same estuary, is broad, ranging from hard sands to thick fluid muds and dense grass beds. We can assume that the same

fundamental hydrodynamic principles hold for the entire
range of bottom types; however, each will have a unique
set of conditions and problems, and thus the entire range
of bottom types should be considered. In many estuaries
the bottom composition is ephemeral, changing with the
seasons, extreme events, or variations in biologic activity. Additional complexity results from the mixture of
organic and inorganic beds, including shell particles
overladen with organic detritus and silts and clays within
seagrass beds. These are but two examples of the diverse
bed mixtures commonly found in estuaries. Although there
have been a few basic studies of these types of problems,
considerable research is needed in order to define the
mechanism of suspended transport in these complex
environments.

One of the more interesting, and better studied, bottom types is the cohesive mud deposit, or fluidized bed.
Many estuaries have zones of thick accumulations of these
mud deposits. The lack of a distinct interface makes it
difficult to document the particulate fluxes in these
deposits. The basic mechansism of erosion (including
resuspension), bed migration, and accumulation rates need
additional study.

The range and complexity of bottom morphologies and
compositions are paralled by the number of hydrodynamic
variables that effect these beds. The transport processes
of suspended particulate in estuaries are controlled by
tidal currents, wind waves, and density currents. That
is, the entire range of forcing functions that determine
the flow field affect the suspended particulates. Particular emphasis should be placed on the role of extreme
events on this system. A sudden change in freshwater
inflow can have a dramatic effect on the suspension of
otherwise stable particulates, Pickral and Odom (1977).
Storm surges, extreme waves, and tides can also result in
a significant increase in the suspended concentrations.
While these responses are generally well recognized, we
have insufficient data and understanding to develop even
crude empirical predictive models of these events.
Extreme temperatures and salinity changes can also have a
significant impact on the rates and zones of suspended
particulate transports. Little data are available to date
to document these effects.

Our understanding of benthic fluxes of a chemical and
biological nature, i.e., the bottom as a source or sink,
is weak. There is a lack of reliable and readily available technology and instrumentation for measuring estua-

rine benthic flux rates that is generally acceptable by
the scientific community. It would seem that an item of
first importance is to develop a technology, or technol-
ogies, which can be tested and proved. Albeit a crude
first attempt at this might be to take the existing procedures and test them together at a number of different
benthic environments.

Such techniques would be useful in studying a variety
of possible benthic flux effects including organic deposition, nutrient regeneration, absorbed metal constituents, bioturbation, benthic algae and aquatic vegetation
effects, and benthic bacterial processes.

On a longer time scale examination and analysis of
bottom cores in estuaries in conjunction with known sedimentation rates is important. Those examples of such
analysis whenever there have been known tracers for dating
have provided some of our most useful information on the
bottom as a sink for certain constituent materials.

The use of numerical modeling and analytic solutions
to hypothesized benthic effects and their comparison with
field data can be a useful procedure for estimating the
importance of certain benthic effects. In this sense the
numerical modeling is used to predict what might happen
under certain conditions rather than as a management prediction tool for what will happen. It is unreasonable to
expect that predictive procedures, either in the form of
a conceptual, analytic, or numerical model, can be carried
out with any certainty for ecosystems analysis in which
the benthic effects are considered to be important and are
not known.

Benthic Remineralization and Recycling Processes

The importance of remineralization processes mediated by
bacteria in organic-rich estuarine and coastal sediments
has been documented through both direct measurements of
the rates of these processes and mass balance budgets of
overlying waters, which include significant benthic flux
terms. For example, a large fraction of phytoplankton
demand for nutrient elements such as nitrogen in coastal
waters is supplied from the sediments. *In situ* flux measurements have confirmed the importance of benthic fluxes
and have examined the sediment-water transport mechanisms
involved.

We have learned a great deal about the rates and sites
of the sedimentary respiration and fermentation reactions
controlling these fluxes during the 1970s. A well-defined

TABLE 4.1 Microbially Mediated Reactions

Reaction	Hydrogen Acceptor	Reduced Product
1. Aerobic respiration	O_2	H_2O
2. Nitrate reduction	NO_3^-	NO_2^-, N_2O, N_2
3. MnO_2 reduction	MnO_2	Mn^{2+}
4. Fe_2O_3 reduction	Fe_2O_3	Fe^{2+}
5. Sulfate reduction	SO_4^{2-}	H_2S
6. Methanogenesis (Fermentation)	CH_3COOH	CH_4

vertical sequence of microbially mediated reactions has
been established for sedimentary environments (Table 4.1)
in which specific bacteria obtain energy for all growth
and maintenance. This sequence appears to arise from
competition in which the organisms capable of greatest
energy yield utilizing available substrates and H accep-
tors dominate. Organic matter accumulating in estuarine
sediments moves downward with time through these biogeo-
chemical zones; however, the most rapid remineralization
occurs near the sediment-water interface.

Because of the relatively high concentrations of dissolved oxygen and sulfate in marine waters and the importance of methanogensis in freshwater sediments, reactions
1, 5, and 6 (Table 4.1) are observed to dominate in organic-rich estuarine sediments. Vertical dimensions of
these zones change in response to the amount and kind of
organic substrates available and to seasonal variations
in temperature. Salt wedge control of dissolved sulfate
distribution leads to upstream dominance of aerobic respiration and methanogenesis. Stable carbon isotope distributions in estuarine sediments indicate the mixing of
apparently less degradable terrestrially derived organic
matter from upstream and surrounding sources with more
degradable planktonic debris downstream.

Recent studies of sediment accumulation and redistribution processes in estuarine and coastal environments
utilizing geochronologic techniques such as excess
^{234}Th, excess ^{210}Pb, $^{239,240}Pu$, and ^{137}Cs provide quantitative means for determining the fate of particle-associ-
ated sedimentary organic debris and, thus, may provide the
necessary framework for understanding the time scales of
degradation processes.

REFERENCES

Aller, R. C. (1980). Diagenetic processes near the sediment-water interface of Long Island Sound. II. Fe and
Mn, *Advan. Geophys. 22*, 351-415.

Carpenter, R., J. T. Bennett, and M. L. Peterson (1981).
^{210}Pb activities in and fluxes to sediments of the
Washington continental slope and shelf, *Geochim.
Cosmochim. Acta 45*, 1155-1172.

Chesapeake Research Consortium (1976). *The Effects of
Tropical Storm Agnes on the Chesapeake Bay System*,
The Johns Hopkins Press, Baltimore, Maryland, 639 pp.

Cutshall, N., W. C. Renfro, D. W. Evans, and V. G. Johnson
(1971). Zinc-65 in Oregon-Washington continental
shelf sediments, in *Radionuclides in Ecosystems*, Proc.
3rd Nat. Symp. on Radioecology, D. J. Nelson, ed.,
CONF-710501-P2, Oak Ridge, Tenn., pp. 694-702.

Elderfield, H., N. Luedtke, R. J. McCafferty, and M.
Bender (1981). Benthic flux studies in Narragansett
Bay, *Am. J. Sci. 281*, 768-787.

Pickral, J. C., and W. E. Odum (1976). Benthic detritus
in a saltmarsh tidal creek, in *Estuarine Processes*,
M. L. Wiley, ed., Estuarine Research Federation,
Academic Press, New York.

Prahl, F. G., and R. Carpenter (1979). The role of zooplankton fecal pellets in the sedimentation of poly-
cyclic aromatic hydrocarbons in Dabob Bay, Wa.,
Geochim. Cosmochim. Acta 43, 1959-1972.

Prahl, F. G., J. T. Bennett, and R. Carpenter (1980). The
early diagenesis of aliphatic hydrocarbons and organic
matter in sedimentary particulates from Dabob Bay,
Wa., *Geochim. Cosmochim. Acta 44*, 1967-1976.

Sholkovitz, E. R. (1978). The flocculation of dissolved
Fe, Mn, Al, Cu, Ni, Co, and Cd during estuarine
mixing, *Earth Planet. Sci. Lett. 41*, 77-86.

Turekian, K. K., J. K. Cochran, L. K. Benninger, and R. C.
Aller (1980). The sources and sinks of nuclides in
Long Island Sound, in *Estuarine Physics and Chemistry:
Studies in Long Island Sound*, B. Saltzman, ed.,
Advances in Geophysics, Vol. 22, Academic Press, New
York.

U.S. Geological Survey (1975). Surface Water Quality of
the United States, *Water Supply Paper 1271*.